Desalination of Seawater

AWWA MANUAL M61

First Edition

American Water Works Association

MANUAL OF WATER SUPPLY PRACTICES — M61, First Edition

Desalination of Seawater

Copyright © 2011 American Water Works Association

All rights reserved. No part of this publication may be reproduced or transmitted in any form or by any means, electronic or mechanical, including photocopy, recording, or any information or retrieval system, except in the form of brief excerpts or quotations for review purposes, without the written permission of the publisher.

Disclaimer
Many of the photographs and illustrative drawings that appear in this book have been furnished through the courtesy of various product distributors and manufacturers. Any mention of trade names, commercial products, or services does not constitute endorsement or recommendation for use by the American Water Works Association or the US Environmental Protection Agency. In no event will AWWA be liable for direct, indirect, special, incidental, or consequential damages arising out of the use of information presented in this book. In particular, AWWA will not be responsible for any costs, including, but not limited to, those incurred as a result of lost revenue. In no event shall AWWA's liability exceed the amount paid for the purchase of this book.

Project Manager/Senior Technical Editor: Melissa Valentine
Production Editor/Cover Design: Cheryl Armstrong
Manuals Specialist: Molly Beach

Library of Congress Cataloging-in-Publication Data
Wetterau, Greg.
  Desalination of seawater / Greg Wetterau, Sandeep Sethi.
    p. cm.
  Includes bibliographical references and index.
  ISBN-13: 978-1-58321-833-4 (alk. paper)
  ISBN-10: 1-58321-833-5 (alk. paper)
 1. Saline water conversion. I. Sethi, Sandeep. II. Title.
 TD479.2.W48 2011
 628.1'67--dc22
                          2011013619

Printed in the United States of America
American Water Works Association
6666 West Quincy Ave.
Denver, CO 80235

Printed on recycled paper

Contents

Chapter 1 Seawater Desalination Overview 1

    Introduction, 1
    Desalination Technologies Overview, 3
    Membrane Separation, 3
    Thermal Evaporation, 6
    Novel Desalination Processes in Development, 9
    References, 13

Chapter 2 Water Quality 15

    Source Water Quality, 15
    Product Water Quality, 17
    Health Concerns, 17
    Product Water Stability, 21
    Irrigation and Industrial Use Concerns, 21
    General Aesthetic Concerns, 23
    References, 24

Chapter 3 Treatment Approaches 27

    Pretreatment, 27
    SWRO Design Parameters, 33
    Disinfection, 36
    Posttreatment, 37
    Energy Recovery, 38
    Corrosion and Materials of Construction, 43
    References, 47

Chapter 4 Environmental Impacts and Mitigation Measures 49

    Introduction, 49
    Source Water Intakes, 50
    Concentrate Discharge, 57
    Management of Desalination Plant Residuals, 69
    Greenhouse Gas Emissions—Impacts and Management, 72
    Noise, Air Pollution, and Traffic, 79
    References, 80

Chapter 5 Cost of Treatment 83

    Introduction, 83
    Summarizing Project Costs, 83
    Construction Costs, 85
    Estimating Capital Costs, 85
    Estimating Operation and Maintenance Costs, 89
    Financing Cost, 93
    Cost of Water, 94
    Summary, 94
    References, 98

Chapter 6　Safety and Security...**99**
    Safety, 99
    Security, 102

Figures

Figure 1-1	Global growth of desalination facilities	2
Figure 1-2	Basic concept of osmosis and reverse osmosis	4
Figure 1-3	Multistage flash distillation	8
Figure 1-4	Multiple effect distillation	8
Figure 1-5	Vapor compression	9
Figure 1-6	Schematic of forward osmosis desalination process	10
Figure 2-1	Sea-surface salinities	16
Figure 2-2	Boston Ivy with tip burn from chloride	20
Figure 2-3	Boron toxicity on camphor	20
Figure 3-1	Projected impact of recovery on power consumption for SWRO	34
Figure 3-2	Projected SWRO feed pressure requirements as a function of influent water temperature for different flux rate and element type	35
Figure 3-3	Projected impact of temperature on SWRO permeate boron	35
Figure 3-4	Pelton wheel generators at Tampa Bay SWRO facility	39
Figure 3-5	Hydraulic turbocharger in an RO system	39
Figure 3-6	ERI™ TurboCharger device (low pressure turbine)	39
Figure 3-7	ERI™ PX energy recovery device flow diagram	40
Figure 3-8	PX™ Pressure exchanger device installation in Sand City, Calif.	40
Figure 3-9	Dual Work Pressure Exchanger flow diagram	41
Figure 3-10	Installation of Flowserve DWEER energy recovery device	41
Figure 3-11	Three center design layout	42
Figure 3-12	Resistance to crevice corrosion	44
Figure 4-1	3.8 MGD intake beach well of a large seawater desalination plant	52
Figure 4-2	Beach well intake system (above-grade completion)	52
Figure 4-3	Beach well intake system (at grade completion)	53
Figure 4-4	Beach well intake system (dual completion)	53
Figure 4-5	Tidal zone (onshore) discharge of the Ashkelon SWRO Plant, Israel	60
Figure 4-6	Perth SWRO Plant discharge configuration	61
Figure 4-7	Perth desalination plant mixing zone	62
Figure 4-8	Perth desalination plant discharge diffuser – rhodamine dye test	63
Figure 4-9	5.5 MGD Santa Barbara seawater desalination plant, California	64
Figure 4-10	Colocation concept for the Carlsbad Seawater Desalination Plant	66
Figure 4-11	Colocation of Tampa Bay Seawater Desalination Plant	66
Figure 4-12	32 MGD Carboneras SWRO plant in Spain	67
Figure 4-13	Carlsbad seawater desalination project	73
Figure 5-1	Seawater RO construction cost	86
Figure 5-2	Seawater RO cost of water	93

Tables

Table 1-1	Operational seawater desalination facilities in the United States	2
Table 2-1	Seawater mineral quality compared to national source waters.	16
Table 2-2	Pathogen reduction requirements for surface waters	19
Table 3-1	Seawater RO pretreatment components for surface seawater sources	30
Table 3-2	Seawater RO treatment advancements for surface seawater sources	31
Table 3-3	Partial list of pretreatment installations in SWRO plants since 1995	32
Table 3-4	Log removal credits for potential treatment processes.	37
Table 3-5	Energy recovery devices (ERD): pros and cons	43
Table 3-6	Galvanic series for alloys in flowing seawater at 4 m/s and 24°C	44
Table 3-7	PREN values for common materials.	46
Table 4-1	Potential impingement/entrainment reduction technologies	56
Table 4-2	Concentrate disposal methods for existing desalination in the U.S. (including brackish RO, NF, and SWRO)	60
Table 4-3	Residuals from seawater desalination processes	70
Table 4-4	Comparison of waste streams from granular media and membrane pretreatment.	71
Table 4-5	Desalination project net GHG emission zero balance	78
Table 4-6	Unit costs of carbon footprint reduction alternatives.	78
Table 5-1	Seawater intake alternatives cost example.	95
Table 5-2	Seawater RO plant capital cost example	96
Table 5-3	Annual Operation and Maintenance Cost Example Treatment Technology: SWRO	97
Table 5-4	Annual Cost of Water Example Treatment Technology: SWRO (with a power plant)	97

Acknowledgments

The first edition of M61 was written through the persistent, dedicated work of the following authors:

G. Wetterau, Chair, Camp, Dresser & McKee Inc., Rancho Cucamonga, Calif.
I. Moch, Sub-committee Chair, I. Moch & Associates, Inc., Wilmington, Del.
V. Frenkel, Kennedy/Jenks Consultants, San Francisco, Calif.
R. Huehmer, CH2M HILL, Englewood, Colo.
H. Hunt, Collector Wells International, Inc., Columbus, Ohio
K. Kiefer, Camp, Dresser, & McKee Inc., Fort Lauderdale, Fla.
T. Pankratz, *Water Desalination Report*, Houston, Texas
S. Sethi, Carollo Engineers, Sarasota, Fla.
G. Silverman, PBSJ Corporation, San Diego, Calif.
S. Trussell, Trussell Technologies, Inc., San Diego, Calif.
L. VandeVenter, AECOM, Wakefield, Mass.
N. Voutchkov, Water Globe Consulting, LLC, Stamford, Conn.

The authors would like to acknowledge the support of the following organizations in preparing this manual:

American Water Works Association, Denver, Colo.
Global Water Intelligence, Oxford, England, United Kingdom

The following individuals provided peer review of this manual. Their knowledge and efforts are gratefully appreciated:

J. Morris, Lead Editor, Metropolitan Water District of Southern California, Los Angeles, Calif.
B. Alspach, Malcolm Pirnie, Inc., Carlsbad, Calif.
K. Kinser, Montgomery Watson Harza, Denver, Colo.
C. Owen, Tampa Bay Water, Tampa, Fla.
G. Silverman, PBSJ Corporation, San Diego, Calif.
L. VandeVenter, AECOM, Wakefield, Mass.
S. Veerapaneni, Black & Veatch Corp., Kansas City, Mo.
N. Voutchkov, Water Globe Consulting, LLC, Stamford, Conn.
G. Wetterau, Camp, Dresser, & McKee, Rancho Cucamonga, Calif.
J. Wong, Brown and Caldwell, Walnut Creek, Calif.

This manual was approved by the AWWA Water Desalting Committee. Members of the committee at the time of approval of this first edition (April 2011) were as follows:

G. Wetterau, Chair, Camp, Dresser & McKee Inc., Rancho Cucamonga, Calif.
S. Veerapaneni, Black & Veatch Corp., Kansas City, Mo.
J. Aguinaldo, Doosan Hydro Technology, Tampa, Fla.
J. Arevalo, Camp, Dresser & McKee Inc., Maitland, Fla.
R.B. Chalmers, Camp, Dresser & McKee Inc., Miami, Fla.
T.D. Chinn, HDR, Inc., Austin, Texas
W. Everest, Malcolm Pirnie, Inc., Irvine, Calif.
A. Franchi, Seal Beach, Calif.
V. Frenkel, Kennedy/Jenks Consultants, San Francisco, Calif.

E. Harrington, AWWA Staff Liaison, Denver, Colo.
Q.C. He, Carollo Engineers, Phoenix, Ariz.
C. Hill, Arcadis, Tampa, Fla.
D. Horne, Virginia Department of Health, Norfolk, Va.
H. Hunt, Ranney Collector Wells International, Inc., Columbus, Ohio
C. Johnson, CH2M HILL, Deerfield Beach, Fla.
D. Khiari, Water Research Foundation, Denver, Colo.
C. Kiefer, Camp, Dresser, & McKee Inc., Fort Lauderdale, Fla.
K. Kinser, Montgomery Watson Harza, Denver, Colo.
J. Loveland, Malcolm Pirnie, Irvine, Calif.
C. Martin, AECOM, Bakersfield, Calif.
S. Mary, Hach Company, Loveland, Colo.
I. Moch, Subcommittee Chair, I. Moch & Associates, Inc., Wilmington, Del.
T. Pankratz, Water Desalination Report, Houston, Texas
R. Pensa, San Francisco Public Utilities Commission, Burlingame, Calif.
R. Rauschmeier, Calif. Public Utilities Commission, San Francisco, Calif.
D. Reel, Black & Veatch Corp., Las Vegas, Nev.
P. Shen, California American Water, San Diego, Calif.
G. Silverman, PBSJ Corporation, San Diego, Calif.
H. Steiman, RW Beck Inc., Framingham, Mass.
S. Talati, San Francisco Public Utilities Commission, Burlingame, Calif.
N. Voutchkov, Water Globe Consulting, LLC, Stamford, Conn.
J. Wong, Brown & Caldwell, Walnut Creek, Calif.

AWWA MANUAL M61

Chapter 1

Seawater Desalination Overview

Sandeep Sethi
Greg Wetterau

INTRODUCTION

As worldwide fresh water supplies become increasingly stressed and world populations continue to grow, seawater desalination has become an increasingly sought-after alternative for new water supply in coastal areas. While three-quarters of the globe is covered with water, less than 0.3 percent is considered a renewable freshwater supply. More than half of the population in the United States lives within 50 miles (80 kilometers) of a coast, so the use of seawater as a source for potable water production is of great interest, especially in areas with stressed and overdrawn freshwater resources. Historically, the high cost of desalination has made it less attractive than freshwater supplies, even where those freshwater supplies were hundreds of miles away (i.e., Southern California).

As desalination becomes more economical, its use for municipal water supply has increased dramatically. Figure 1-1 shows that the worldwide desalination capacity more than doubled between 2002 and 2010. In the United States, most desalination facilities treat brackish water or are membrane softening plants; however, seawater desalination plants currently outnumber brackish water plants by 60 percent worldwide (GWI 2009).

Table 1 lists some of the more than two dozen seawater desalination plants built and operated in the United States. The majority of these facilities are industrial with a capacity of less than 1 million gallons per day (mgd) or 3.8 megaliters per day (MLD). In addition, a number of these plants are used intermittently because of the high cost of operation or problems experienced during operation. As coastal municipalities in the United States begin to consider implementing larger seawater facilities, it is essential to ensure that these are constructed and operated in an efficient and reliable fashion without adversely impacting fragile coastal environments. Large capacity, highly efficient seawater desalination facilities have been successfully implemented within the last five years in Australia, Singapore, Spain, and several countries in the Middle East. In the United States, there are currently more than two dozen new seawater projects in various stages of development, primarily in California, Texas, and Florida.

The purpose of this manual of practice is to identify lessons learned from recent studies and seawater desalination projects around the world, and to use these to provide guidance for seawater desalination facilities that are reliable, economical, and environmentally sound.

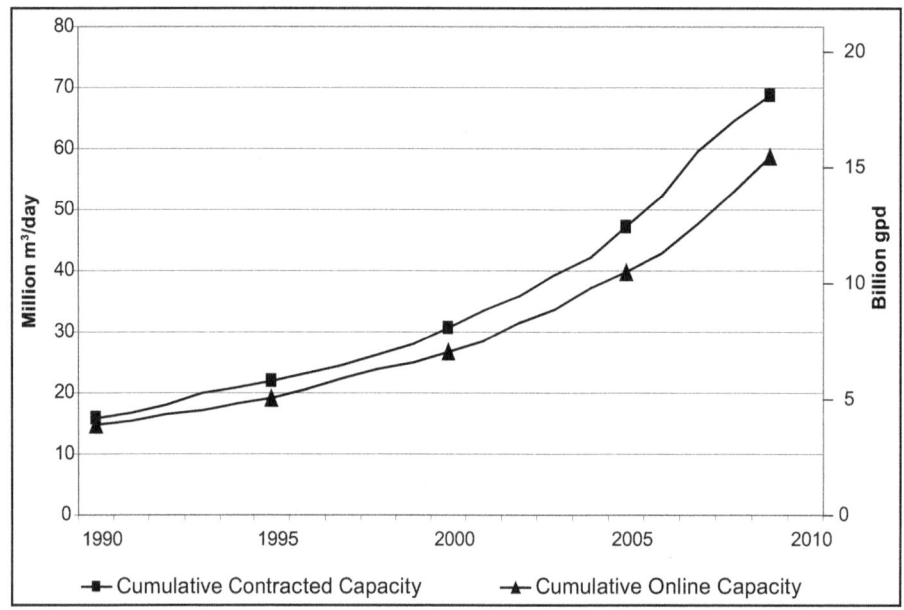

Source: Global Water Intelligence 2010, data reproduced from DeSalData.com/Desalination Markets

Figure 1-1   Global growth of desalination facilities

Table 1-1   Operational seawater desalination facilities in the United States

Diablo Canyon, CA (0.6 mgd or 2.3 MLD)	Tampa, FL (25 mgd or 95 MLD)
Gaviota, CA (0.4 mgd or 1.5 MLD)	Stock Island, FL (2 mgd or 8 MLD)
Morro Bay, CA (0.6 mgd or 2.3 MLD)	Marathon, FL (1 mgd or 4 MLD)
Moss Landing, CA (0.5 mgd or 1.9 MLD)	Kauai, HI (0.2 mgd or 0.8 MLD)
Monterey Bay Aquarium, CA (0.04 mgd or 0.15 MLD)	Swansea, MA (2 mgd or 8 MLD)
Sand City, CA (0.3 mgd or 1 MLD)	Brockton, MA (5 mgd or 19 MLD)
Avalon, CA (0.1 mgd or 0.4 MLD)	

Courtesy of Greg Wetterau

DESALINATION TECHNOLOGIES OVERVIEW

Desalination processes can be divided into two broad categories: membrane separation and thermal evaporation. Membrane-based desalination processes typically employ mechanical pressure, electrical potential, or a concentration gradient as the driving force across a semi-permeable membrane barrier to achieve physical separation. Thermal desalination processes employ heat to evaporate the water from a salt solution, and the water vapor is then condensed and recovered.

Thermal technologies were the only options available for seawater desalination until reverse osmosis (RO) membranes were developed in the early 1960s. Since then, RO membrane processes have steadily been improved, and the efficiency has increased to the point that they are now the technology of choice for most seawater desalination applications. An exception to this is the Middle East, where low energy costs allow for thermal desalination to remain relatively competitive.

Besides the established desalination technologies, there are several newer technologies that are nearing commercialization or undergoing active research and development. A discussion of the established membrane and thermal technologies is presented first in this manual, followed by a brief discussion of developing technologies. The remaining chapters in this manual focus on pressure-driven membrane applications, as this presently has the most applicability to seawater desalination in the United States.

MEMBRANE SEPARATION

Membrane desalination technologies have been designed around the ability of semi-permeable membranes to selectively permit or minimize the passage of certain ions. Three fundamental driving forces can be used in membrane desalination systems including pressure, electric potential, and concentration gradient. RO and nanofiltration (NF) are pressure driven processes. Electrodialysis (ED) and electrodialysis reversal (EDR) are electric potential driven processes. Forward osmosis (FO) is a concentration-driven process.

Membrane-based seawater desalination processes have typically applied only RO. Although NF and ED/EDR are also mature technologies and can be used for desalination, ED/EDR are typically not cost competitive for desalination of seawater (Amjad 1993), and NF is not ordinarily considered for seawater desalination for potable water production. However, a novel approach employing two-pass (NF) configuration has been developed and tested for seawater desalination by the Long Beach Water Department in California. Similarly, FO is a developing technology and has not yet been commercialized for large-scale applications.

Reverse Osmosis (RO)

Desalination through RO is a well-established and nonproprietary unit process that currently represents the state-of-the-art of desalination technology for a number of reasons. In addition to the ability to reject a variety of contaminants, RO treatment generally has lower energy consumption, lower feed water flows, and no thermal impacts in the concentrate discharge in comparison to thermal desalination processes. Improvements in membranes and energy recovery devices used for seawater RO (SWRO) have improved the overall process efficiency thereby lowering the costs associated with treatment.

Reverse osmosis is based on overcoming the natural phenomenon of osmotic pressure, which occurs when a semi-permeable membrane separates two solutions with different concentrations of ions. The osmotic pressure created by the concentration gradient drives the flow of water from the dilute solution to the concentrated solution, until chemical equilibrium is established. The flow of water can be reversed with the application of an external hydraulic force (pressure) if this force is greater than the osmotic pressure. Figure 1-2 illustrates the basic concepts of osmosis and reverse osmosis.

4 DESALINATION OF SEAWATER

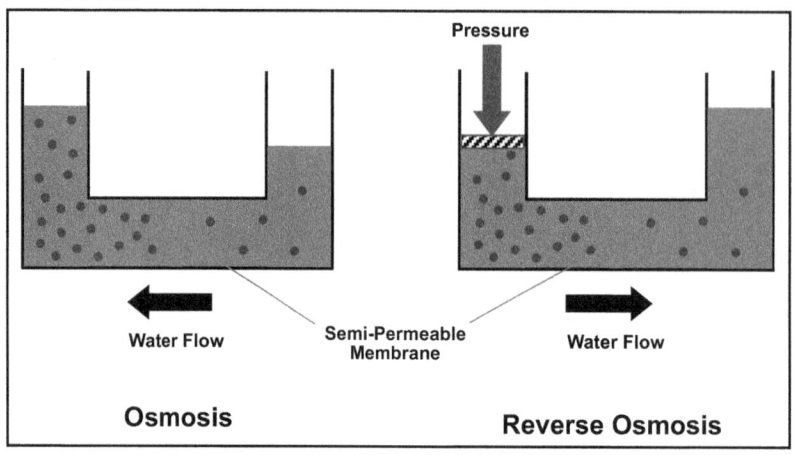

Courtesy of Sandeep Sethi

Figure 1-2 Basic concept of osmosis and reverse osmosis

RO membranes are designed to retain salts and low-molecular weight solutes while allowing water to pass through. The original asymmetric cellulose acetate (CA) membranes, developed in the 1960s, were less permeable than modern thin-film composite (TFC) membranes and required a higher driving pressure, in excess of 1200 pounds per square inch (psi) or 8.3 megapascals (MPa) for seawater at typical operating fluxes. Additionally, the ability of CA membranes to reject salts was originally less than current materials.

Cellulose acetate membranes utilized an asymmetric structure while the TFC contained multiple layers made from different materials. In the asymmetric configuration, the membrane consists of the same material throughout with a dense layer at the top and porous layer beneath. In contrast, the TFC membrane consists of a thin but dense layer of one material over a porous support consisting of a different material.

Currently, there are a variety of modified and improved blends of CA membranes available to the desalination industry, but these membranes are rarely used in large-scale desalination applications. CA membranes can tolerate continuous exposure to low concentrations of chlorine (0.1 to 0.5 mg/L at 25°C), which is an advantage for biofouling control in seawater applications. They are, however, susceptible to hydrolysis, which compromises the membrane's salt rejection performance. Hydrolysis of CA membranes is accelerated if the operating pH is less than approximately 4 or greater than approximately 7 and temperatures are greater than 30°C (Mallevialle et al. 1996). Therefore, pH depression into this range is needed for seawater desalination with CA membranes.

The development of TFC membranes provided greater salt rejection and higher water production per unit membrane area. TFC membranes are made by combining a thin, dense membrane film with a porous underlying material that provides structural support. The thin film typically consists of aromatic polyamide (PA) and the bottom support layer is typically polysulfone. Most of the solute rejection occurs at the thin nonporous film, and its small thickness can significantly reduce the pressure required to drive water through it in comparison to CA membranes. TFC membranes are stable over a broad pH range (2-11) and can withstand temperatures as high as 45°C. However, unlike the CA membranes, they are susceptible to degradation by strong oxidants such as free chlorine. Although the degradation rate caused by free chlorine is a function of pH, membrane materials generally deteriorate upon exposure to chlorine (sometimes catastrophically).

High pressures are required to overcome the osmotic pressure of the salts and minerals, and the resistance from the membrane material, and other associated system losses.

SWRO membranes are typically operated at feed pressures of approximately 800 to 1,000 psi (5.5 to 6.9 MPa). RO membranes are capable of rejecting contaminants as small as 0.1 nm; however, the process of water transfer is mostly diffusion controlled rather than convection controlled as with microfiltration and ultrafiltration. In addition to the effects of the major ion matrices, mass transfer of ions through RO membranes is also impacted by broader water quality characteristics, such as temperature and pH.

The amount of water recovered using SWRO membranes ranges from 35 to 60 percent, and commercially available SWRO membranes typically reject 99.5 to 99.8 percent of the total dissolved solids (TDS) in the feed water. However, removal of a few constituents, such as boron, is sometimes not as great as might be required (see Chapter 2). If product water goals are not met, additional treatment may consist of two-pass RO, in which a portion or all of the permeate produced in the first pass is treated again in a second pass. New membranes with improved boron rejection are currently being developed by SWRO manufacturers to avoid the need for two-pass treatment; however, other water quality goals besides boron may also impact the need for a two-pass system. An optimized SWRO design will therefore depend on the feed water quality, system operating conditions, and specific finished water quality requirements.

Because membrane processes are based on physical separation, they do not require thermal energy to vaporize the water (with the exception being membrane distillation, discussed later in this chapter). As a result, the energy consumption for treatment components of an SWRO plant typically falls in a range of 10 to 20 kilowatt-hours (kWh)/1,000 gallons (2.6 to 5.3 kWh/m^3). In comparison, total energy used for thermal desalination treatment processes can range from 10 to 40 kWh/1,000 gallons (2.6 to 10.6 kWh/m^3), depending on the unit processes.

Energy recovery devices are increasingly used in SWRO applications. These devices can recover from 25 to over 45 percent of input energy for SWRO. Examples of such devices as presented in Chapter 3, include Pelton wheels, work exchangers, pressure exchangers, and hydraulic turbo-exchangers.

One of the greatest challenges for membrane desalination processes is fouling and scaling of the membranes. Fouling can occur as a result of inadequate pretreatment or measures for reduction of particulate, colloidal, or organic matter to tolerable levels, or biological growth in the membrane pressure vessels. Scaling results from precipitation of sparingly soluble salts in the system and tends to be less of a concern in seawater desalination than in brackish water systems, which run at higher recoveries. Compounds such as calcium carbonate, calcium sulfate, silicate, barium sulfate, and strontium sulfate in the feed water may, however, contribute to limiting the recovery of the RO process. Acid or scale inhibitors (also known as *antiscalants*) may be added to reduce alkalinity and prevent formation of scale, allowing for higher recovery than otherwise possible. As a result of high levels of particulates and the generally aerobic state of seawater, SWRO plants require comprehensive pretreatment and chemical conditioning of the feedwater for successful operation.

Nanofiltration (NF)

Nanofiltration is typically used to soften water and remove disinfection by-products (DBP) precursors such as dissolved organic matter. NF is typically not used for seawater desalination, although unique configurations of two-pass NF have been successfully used to desalinate seawater.

Nanofiltration uses semi-permeable membranes and a driving force of hydraulic pressure; however, in comparisons to RO, NF membranes typically have a higher molecular weight cut-off (MWCO). NF membranes remove a high percentage (90 to 98 percent) of divalent ions (i.e., those associated with hardness) but removal of monovalent ions is somewhat limited (typically 60 to 85 percent).

Because a higher concentration of monovalent ions can pass through the NF membrane, the osmotic pressure is lower compared to RO. This, combined with a more permeable membrane skin layer, reduces the hydraulic pressure requirements to 500 to 700 psi (3.4 to 4.8 MPa) for seawater applications. Recognizing these advantages, the Long Beach Water Department (California, United States) has developed and patented an innovative two-pass nanofiltration method for the desalination of seawater. This will be discussed further in Chapter 3.

Electrodialysis (ED) / Electrodialysis Reversal (EDR)

ED and EDR processes use ion-selective membranes and an electrical potential as a driving force to separate charged species from water. Pressure driven systems (RO and NF) selectively pass water through a membrane and retain dissolved salts in the concentrate. In contrast, ED and EDR use an electrical potential to draw dissolved ions through a set of membranes (cations to one side, anions to the other), while the deionized water passes between the membranes and is ultimately recovered.

An electrodialysis stack consisting of alternating layers of cationic and anionic ion-selective flat-sheet membranes creates channels of desalted product water and concentrated reject water. Cations migrate to the cathode and anions migrate to the anode while cation-selective membranes allow only cations to pass and anion-selective membranes allow only anions to pass. The net effect is to remove the salt from every other cell.

A modification of the ED process, EDR, periodically reverses the polarity of the applied electrical potential on the stack to minimize the effects of inorganic scaling and fouling by switching product channels into concentrate channels and vice versa. This allows the EDR system to operate at higher recoveries compared to ED.

ED/EDR processes are typically not used for seawater desalination because with higher salinities, the ED/EDR process generally becomes less efficient than other membrane-based desalination technologies. Additionally, bacteria, nonionic constituents, and residual turbidity are not affected by this process and therefore remain in the product water, requiring additional treatment before drinking water standards are met. Because ED/EDR are not typically used for seawater desalination, these processes will not be discussed further in this manual.

THERMAL EVAPORATION

Thermal desalination technologies work by evaporating water from a saline solution and then condensing the vapor (steam) to produce distilled water. All large-scale thermal processes involve heating water to its boiling temperature to produce the maximum amount of water vapor. The pressure of the system is typically decreased so that the temperature required for boiling is reduced. Commercially available distillation systems are designed to allow for "multiple boiling" in a series of vessels that operate at successively lower temperatures and pressures.

Thermal technologies that are used for desalination include multistage flash (MSF), multiple effect distillation (MED), and vapor compression (VC). MSF and MED systems typically use direct heat exchange from steam as the energy source for evaporation, while VC systems use the heat from the compression of the vapor as the energy source for evaporation. Thermal processes can produce water with very low salt concentrations (TDS levels of 10 mg/L or less) from TDS levels as high as 60,000-70,000 mg/L TDS; however, there are limitations associated with distillation processes for seawater desalination.

One of the most significant limitations of thermal technologies is the energy requirement of the vaporization step. High levels of salts result in boiling point elevation, and the energy required to vaporize seawater ranges from around 25 to 100 kWh/1000 gal of fresh water produced (Wade 2001). It should be noted that these thermal energy requirements are in addition to the electrical energy required for the other aspects of the process. Often, large distillation plants are coupled with steam or gas turbine power plants, making use of low grade heat to reduce power input requirements. Thermal technologies are more commonly used in the Middle East, where energy costs are relatively low, the large land requirements are not cost prohibitive, and ecological permitting requirements are less stringent. There has long been interest in using solar energy as a source of heat for accomplishing the evaporation in distillation, but suitable technologies for a large-scale project are not yet available.

Operational issues for thermal desalination include corrosion and scaling. Because seawater is highly corrosive in nature, special alloys, such as cupronickel alloys, aluminum, and titanium, are used most commonly in desalination with distillation processes. These special alloys contribute significantly to the capital cost of a distillation plant, particularly with the large surface area required for efficient distillation. The scaling of sparingly soluble salts at elevated temperatures on the inner walls of pipes and equipment is another operational issue that reduces the heat transfer efficiency of the heat exchangers, increasing the overall energy required for distillation. Also, additional permitting concerns may arise because concentrate discharged from a thermal distillation process has a higher temperature than the ambient water in the discharge location. While the cost of thermal desalination is often considerably higher than RO, very little pretreatment is required ahead of thermal processes, and the product water quality is extremely high (less than 10 mg/L TDS), avoiding the need for additional treatment to address boron, chloride, or bromide concerns.

Multistage Flash Distillation (MSF)

MSF accounts for the greatest installed thermal distillation capacity worldwide. In the MSF process, water is heated in a series of stages, each with successively lower pressures and temperatures. Typically, MSF plants can contain from 15 to 25 stages. Vapor generation or boiling caused by reduction in pressure is known as *flashing* (illustrated in Figure 1-3). As the water enters each stage through a pressure-reducing nozzle, a portion of the water is flashed to form vapor. In turn, the flashed water condenses on the outside of the condenser tubes and is collected in trays. As the vapor condenses, the latent heat is used to preheat the seawater that is being returned to the main heater, where it will receive additional heat before being introduced to the first flashing stage. The condensate collected in each stage forms the product, and the whole process is driven by a subatmospheric pressure gradient through the stages.

Evaporation or flashing of a small portion of the feed continues in each successive stage at a lower pressure. The MSF process generates and condenses vapor in the same stage or effect. The range of recoveries for conventional MSF desalination processes is limited to about 10 to 30 percent for seawater desalination.

Multiple Effect Distillation (MED)

The MED process, like the MSF process, uses multiple vessels (or effects) arranged in series with reduced ambient pressure in each subsequent effect. Typically, 8 to 16 effects are used in MED to minimize the energy consumption. The feed water is distributed on the outside of the evaporator tubes in a thin film (see illustration in Figure 1-4) to promote rapid boiling and evaporation. Steam is condensed on the colder inside surface. Vapor produced by evaporation is condensed in a way that uses the heat of vaporization to heat the remaining saline solution at a lower temperature and pressure in each succeeding effect, allowing water to undergo multiple boiling without supplying any additional heat after the first effect. Thus, the vapor produced in each effect is used to heat the feed water in the next effect. This not only reduces the energy required for distillation but also the overall electrical power consumption. As a result, energy costs for operating an MED plant are lower than that of an MSF plant.

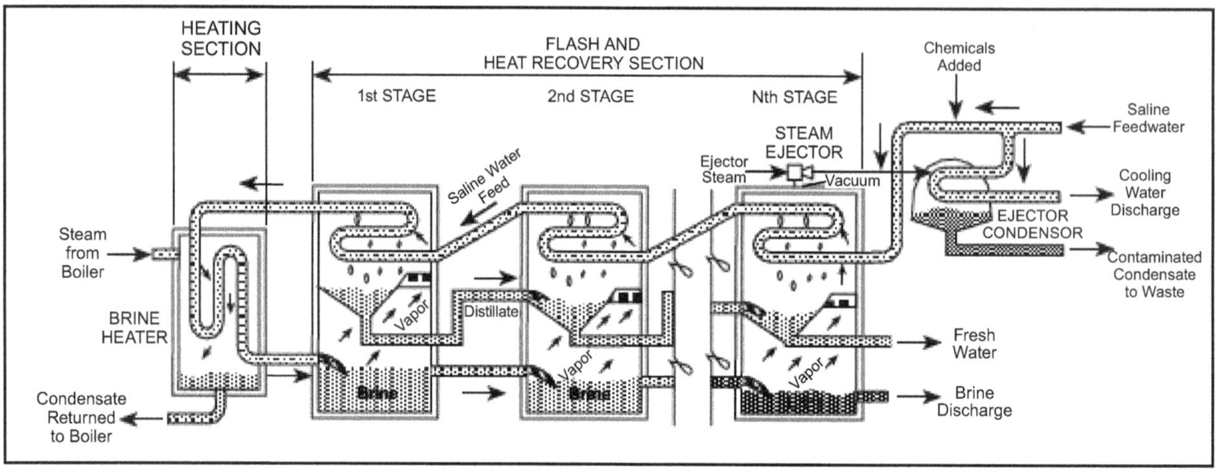

8 DESALINATION OF SEAWATER

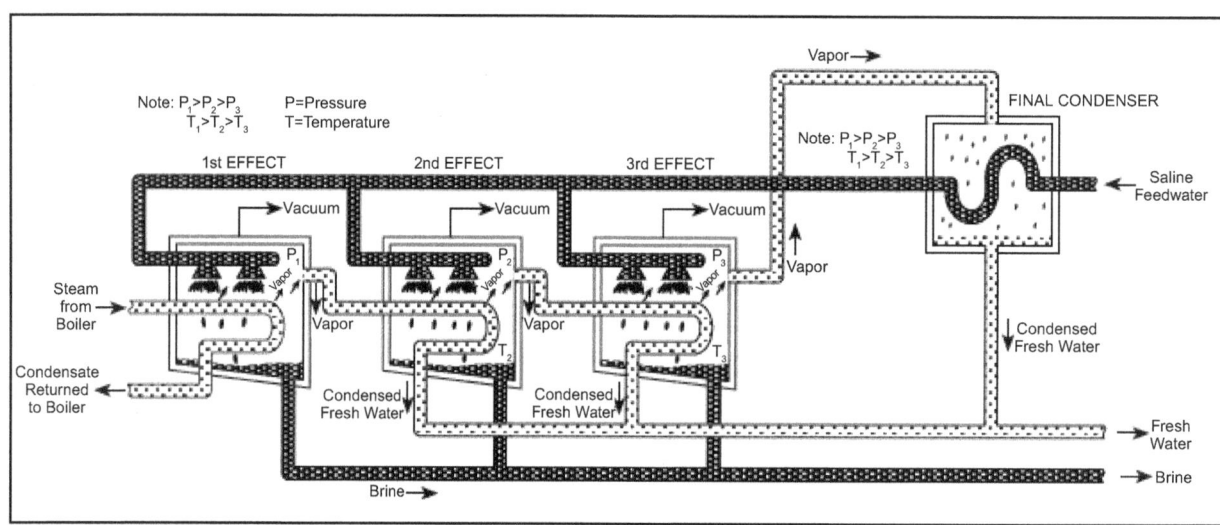

Source: Buros 2000 Water Desalting Planning Guide. Reprinted with permission of John Wiley & Sons, Inc.

Figure 1-3 Multistage flash distillation

Source: Buros 2000 Water Desalting Planning Guide. Reprinted with permission of John Wiley & Sons, Inc.

Figure 1-4 Multiple effect distillation

The steam generated in the final effect is typically at a pressure and temperature too low to be of further use. MED systems normally condense this steam using an external cooling source to remove the heat of condensation.

Energy is required in an MED system as follows: (1) to create steam of sufficient pressure to drive evaporation in the first stage; (2) to power vacuum systems to reduce the boiling pressure in the downstream effects (if operated at low temperatures); (3) to pump influent water through the heat exchangers to the evaporator(s), to recirculate the concentrate within each evaporator stage, and to pump the condensate and concentrate through the heat recovery prior to exiting the system; (4) cooling water to condense the steam from the final stage. Energy efficiencies may be gained via the combination of the evaporator systems with available low-pressure or waste steam/heat sources or by the addition of efficiency enhancement devices to the conventional MED system. The range of recoveries for conventional MED is limited to approximately 20 to 35 percent for seawater desalination.

Vapor Compression (VC)

Heat for evaporation in VC systems is provided by one of two approaches: mechanical vapor compression (MVC) or thermo vapor compression (TVC); an illustration of the former is provided in Figure 1-5. MVC systems use electricity while TVC systems use high-pressure steam to compress the water vapor created from distillation to higher pressure and temperature, so that it can be returned to the evaporator and used as a heat source. The vapor compression process is well established and is used for seawater desalination as well as treating RO concentrate for residuals management. Vapor compression systems typically have recoveries in the range of 40 to 50 percent for seawater desalination.

NOVEL DESALINATION PROCESSES IN DEVELOPMENT

Forward Osmosis

As in the case of RO and NF, FO employs a semi-permeable membrane to separate water from a saline solution; however, instead of using external hydraulic pressure to create the driving force for water transport through the membrane, the FO process employs a natural pressure gradient provided by a higher salinity "draw" solution (such as ammonium carbonate or specially prepared magnetic nanoparticles). The higher osmotic pressure of the draw solution causes water to move toward it through a membrane. Freshwater is then separated from the draw solution using an additional separation process, which can vary depending on the nature of the draw solute. The separated draw solutes are either recovered and reused in the FO process or discharged.

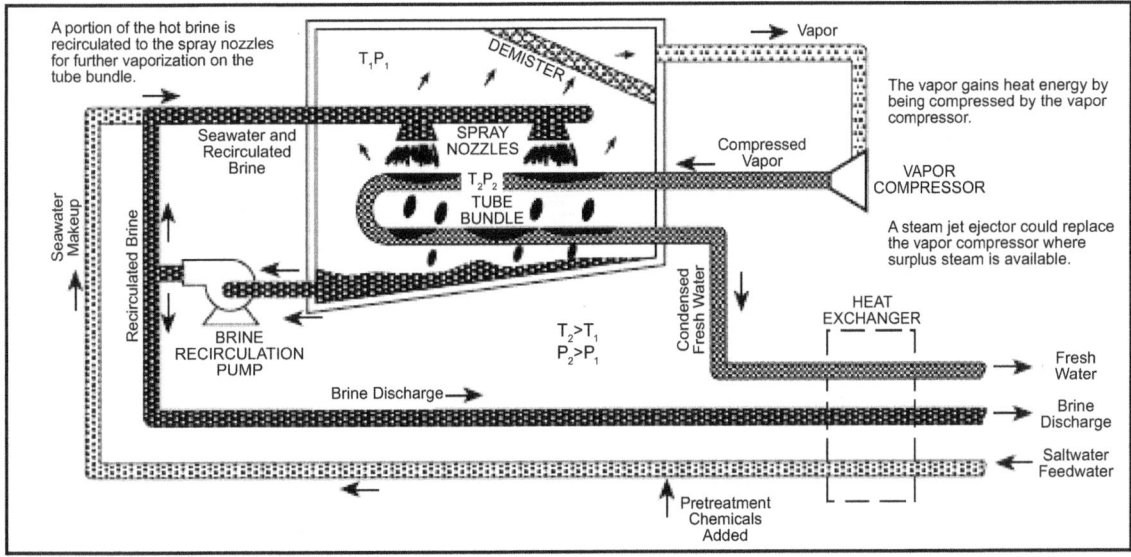

Source: Buros 2000 Water Desalting Planning Guide. Reprinted with permission of John Wiley & Sons, Inc.

Figure 1-5 Vapor compression

Osmotic driving forces in FO can be significantly greater than hydraulic driving forces used in RO. The use of a suitable draw solution with very high osmotic pressure driving forces can be used in principle to generate high water fluxes and recoveries. The FO process, once fully developed and commercialized, is expected to have potential advantages in terms of relatively low fouling potential, low energy consumption, and simplicity. Identification of appropriate draw solutions and development of efficient membranes are two of the most pressing challenges for FO. An effective draw solute should have the following characteristics:

- High osmotic efficiency, meaning that it has to be highly soluble in water and have a low molecular weight in order to generate a high osmotic pressure.
- Nontoxic as trace amounts may be present in the product water.
- Chemical compatibility of the membrane.
- Easy and economical separation from recovered water.

Commercial RO membranes are not suited for the FO process because of relatively low product water fluxes. This low water flux is due mainly to internal concentration polarization within the porous support layer of the membrane, which alters the effective driving force across the active layer of the membrane, thereby limiting water flux. One of the important tasks for future research is the development of a semi-permeable FO membrane having high salt rejection and minimal internal concentration polarization to realize higher product water fluxes. Figure 1-6 illustrates the general process used in forward osmosis.

FO technology is still in development. Bench-scale FO units have been built and operated at Yale University laboratory (McCutcheon and Elimelech 2006) supported by Office of Naval Research (Award No. N000140311004). The draw solution employed in this study consists of highly concentrated ammonium carbonate, prepared by mixing ammonia and carbon dioxide gases. Upon heating (to approximately 60°C), ammonium carbonate decomposes back into ammonia and carbon dioxide gases, leaving behind the desalinated water. This separation needs to be essentially complete because of limits on ammonia in drinking water. Thermal recovery of ammonia and carbon dioxide from the draw solution requires energy, so the process may be suited for applications where low-grade heat is available.

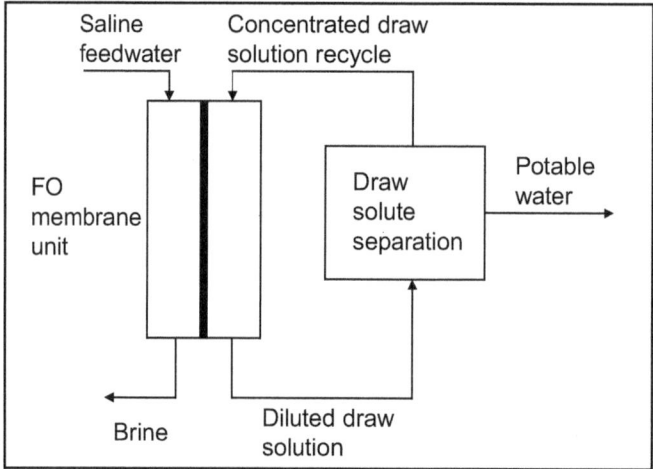

Source: Water Research Foundation 2009. Reprinted with permission.

Figure 1-6 Schematic of forward osmosis desalination process

FO has also been studied for use as pretreatment for RO in several novel applications including water reclamation and nutrient recovery (Cath et al. 2005, Holloway et al. 2007). Several hybrid processes have also been developed including osmotic dilution prior to desalination, which reportedly can significantly reduce the energy demand of desalination in some applications (Lundin 2009). Finally, FO has been proposed as a method of energy generation in blending highly concentrated desalination brine with fresh wastewater flows prior to discharging into the ocean. Such an approach would employ two waste products to produce electrical power and reduce the overall energy footprint of the desalination facility.

Membrane Distillation

Membrane distillation (MD) is a hybrid process using principles of both membrane separation and thermal distillation. MD involves evaporation of water from a saline solution and transport of the water vapor through the pores of a hydrophobic membrane. The membrane allows water vapor to pass through but prevents the solution from passing through. Water vapor is transported across the membrane in response to a change in partial pressure across the membrane because of a thermal gradient. The clean vapor is subsequently carried away from the membrane and condensed as pure water either within the membrane package or in a separate condenser system.

MD differs from pressure-driven membrane technologies in that, rather than applying pressure to force liquid through a membrane, the driving force for desalination is the difference in vapor pressure of the liquid across the membrane. Increasing the temperature of the liquid increases the vapor pressure and results in increased membrane penetration rate. The efficiency of an MD process largely depends on the feed water quality, system design, and heat recovery from the permeate stream. MD has been reported to run at relatively low temperature (approximately 70°C) and thus can utilize waste heat or low-grade heat sources. The energy source for feedwater heating and/or for a vacuum system to sweep away the vapor may be low-grade thermal energy such as supplied by low-pressure steam, waste heat, solar energy, or geothermal energy.

Potential advantages of MD are the ability to use low-grade heat, minimal pretreatment needs, and negligible scaling or precipitation concerns. Challenges include need of waste heat for economic feasibility, membrane fouling, and membrane degradation due to loss of hydrophobicity. MD technology is currently in the development and demonstration phase.

A variety of arrangements and configurations can be used to induce the vapor through the membrane, collecting and condensing it as product water. Common to all concepts is that the feedwater directly contacts the membrane. Condensation can be achieved using one of four process configurations:

1. Air-Gap Membrane Distillation.   This configuration, which is the most common and most versatile arrangement, provides an air gap after the membrane, followed by a cool surface for condensation to occur.

2. Direct-Contact Membrane Distillation.   The cool condensing solution (pure water) directly contacts the membrane and condenses the vapor as it passes through the membrane, where the coolant liquid typically flows countercurrent to the feed water.

3. Sweep-Gas Membrane Distillation.   A sweep gas pulls the water vapor out of the membrane gap for subsequent condensation outside of the membrane package.

4. Vacuum Membrane Distillation.   Vacuum is applied to the membrane space to pull the water vapor out of the system.

Freeze/Thaw

The freeze/thaw approach to desalination is similar to thermal desalination in the fundamental concept of relying on a phase change to achieve separation. In the case of freeze/thaw, the phase change is from liquid to solid. Ice crystals exclude salt from their structure and the salt is then able to be separated as a brine from the ice. A key aspect of the process is that the energy required for the phase change from water to ice is less than one-seventh the energy that is required for the phase change from water to vapor (however, practical thermal desalination processes such as MSF and MED use much less energy than the heat of evaporation due to the use of multiple effects or stages as previously discussed). Challenges with freeze desalination include implementing the proper washing and separation of the crystals without premature melting and/or recontamination with the excluded salt. Different configurations of freeze/thaw desalination systems that have been developed include direct freezing, indirect freezing, and absorption (AWWA 2004).

Capacitive Deionization (CDI)

Carbon aerogel is an ideal electrode material because of its high electrical conductivity, high specific surface area, and controllable pore size distribution (Yang et al. 2001, Ying et al. 2002). The Lawrence Livermore National Laboratory (LLNL) began its research into carbon aerogels and capacitive deionization technology (CDT) in the late 1980s. LLNL developed and optimized carbon aerogel materials, which multiplied the effective surface area of the deionization electrodes by a factor of 60,000, dramatically improving their capacity to attract and hold charged water constituents.

In the process of capacitive deionization (CDI), the high salinity solution flows between the electrode pairs and ions are adsorbed onto the surface of the porous electrodes by applying a low voltage electric field thereby producing deionized water. The major mechanisms related to the removal of charged constituents during CDI are physisorption, chemisorption, electrodeposition, and/or electrophoresis. Unlike ion exchange, no additional chemicals are required for regeneration of the electrosorbent in this system. Adsorbed ions are desorbed from the surface of the electrodes by eliminating the electric field, resulting in the regeneration of the electrodes. The efficiency of CDI strongly depends on the surface property of electrodes, such as their surface area and adsorption properties.

CDI systems exhibit several advantages: a simple, modular, plate-and-frame construction, and low energy requirement. However, current challenges include the limited adsorption capacity of carbon aerogel electrodes, slow kinetics of transport of ions into and out of the highly porous electrodes, relatively high costs of the CDI modules, and the fouling potential of the aerogel surface caused by natural organic matter. CDI is still in the development stage with on-going bench and pilot tests.

Supercritical Desalination (SCD)

Recent research by the Wetsus, Centre of Excellence for Sustainable Water Technology in the Netherlands (Ingo et al. 2009) has looked at the use of supercritical conditions in water to promote desalination. Supercritical conditions are achieved at elevated temperatures and pressures, where the liquid and gas form of water both exist and are indistinguishable from each other. Supercritical water is an extremely poor solvent of inorganic salts, allowing dissolved salts to be removed through precipitation, producing a solid or near solid waste stream and a purified product. The purity of the product depends on the temperature and pressure of the supercritical fluid; however, SCD has been proposed as a pretreatment step for RO in seawater desalination, where SCD is used to produce a water with TDS levels close to 3,500 mg/L using a pressure of 22 MPa (3,200 psi) and temperature of approximately 350°C. Brackish water reverse osmosis (BWRO) membranes could then be used for further desalination, with the theoretical energy requirements being comparable with conventional SWRO, but with higher recoveries on the order of 70 to 80 percent. SCD technologies are still in development and have not been tested outside of the laboratory; however, these technologies may play a role in future desalination applications as they offer unique opportunities not currently available with either membrane- or evaporation-based technologies.

REFERENCES

American Water Works Association. 2004. *Water Desalting Planning Guide for Water Utilities*. New York, N.Y.: John Wiley & Sons.

Amjad, Z. 1993. *Reverse Osmosis: Membrane Technology*. Water, Chemistry, and Industrial Applications, Van Nostrand Encyclopedia of Chemistry. Reinhold, New York: John Wiley & Sons.

Cath, T.Y., and A.E. Childress. 2005. Membrane contactor processes for wastewater reclamation in space. II. Combined direct osmosis, osmotic distillation, and membrane distillation for treatment of metabolic wastewater, *Journal of Membrane Science* 257: 111-119.

Global Water Intelligence. 2010. DeSalData.com. Oxford, England.

Holloway, R.W., A. E. Childress, K.E. Dennett, and T.Y. Cath. 2007. Forward osmosis for concentration of centrate from anaerobic digester, *Water Research* 41: 4005-4014.

Ingo, L., S.J. Metz, G. Rexwinkel, and G.F. Versteeg. 2009. Desalination via Supercritical Water: A New Approach for Desalination without Brine Streams. Proceedings of the International Desalination Association 2009 World Congress, November 7-12, 2009, Dubai, UAE.

Lundin, C., J.E. Drewes, T.Y. Cath. 2009. A Novel Hybrid Forward Osmosis—Reverse Osmosis Process for Water Purification and Reuse, Using Impaired Water and Saline Water. Proceedings of the American Water Works Association 2009 Membrane Technology Conference, March 15–18, 2009, Memphis, Tennessee.

Mallevialle, J., P.E. Odendaal, and M.R. Wiesner. 1996. *Water Treatment Membrane Processes*. Denver, Colo.: AWWA and AWWA Research Foundation.

McCutcheon, J.R. and M. Elimelech. 2006. Influence of Concentrative and Dilutive Internal Concentration Polarization on Flux Behavior in Forward Osmosis. *Journal of Membrane Science*, 284(1-2):237-247.

Sethi, S., Walker, S., Xu, P. and Drewes, J.E. 2009. Desalination Product Water Recovery and Concentrate Volume Minimization. Water Research Foundation.

Wade, N.M. 2001. Distillation plant development and cost update. *Desalination* 136:3–12.

Yang K-L, T-Y Ying, S. Yiacourmi, C. Tsouris, and E.S. Vittoratos. 2001. Electrosorption of ions from aqueous solutions by carbon aerogel: an electrical double layer model. *Langmuir* 17: 1961-1969.

Ying, T-Y., K-L Yang, S. Yiacoumi, and C. Tsouris. 2002. Electrosorption of Ions from Aqueous Solutions by Nanostructured Carbon Aerogel. *Journal of Colloids and Interface Science* 250, 18-27.

Chapter 2

Water Quality

Greg Wetterau
Shane Trussell

SOURCE WATER QUALITY

While seawater quality in the open ocean does not vary more than 10 percent over time and location, large variability can be seen in partially isolated seawater bodies, such as bays, estuaries, and seas, where influences from freshwater flows and evaporation have considerable impact on the water quality patterns. Figure 2-1 shows the worldwide surface salinity of oceans, as modeled by the Ocean Climate Laboratory of the National Oceanographic Data Center (World Ocean Atlas 2005). The figure presents salinity in Practical Salinity Units (psu), with one psu equal to 1,000 milligrams per liter (mg/L). While higher than average salinities are common in areas receiving little rainfall, such as the Red Sea, the Mediterranean, and the Arabian Gulf, seawater bodies surrounding the United States are typically average or below average in salinity. Considerably lower salinities are also seen in the U.S. in partially closed water bodies receiving high contributions from freshwater flows, such as the San Francisco Bay, Tampa Bay, and Chesapeake Bay, where seawater desalination facilities are either currently operational or are being actively evaluated. Extensive monitoring of seawater quality was first documented by William Dittmar in 1884, after four years of water quality sampling over nearly 70,000 nautical miles. The extremely high concentrations of dissolved minerals, many of which are orders of magnitude higher than typical fresh water supplies (see Table 2-1), demonstrate the challenge of treating seawater to a quality acceptable for potable water use.

16 DESALINATION OF SEAWATER

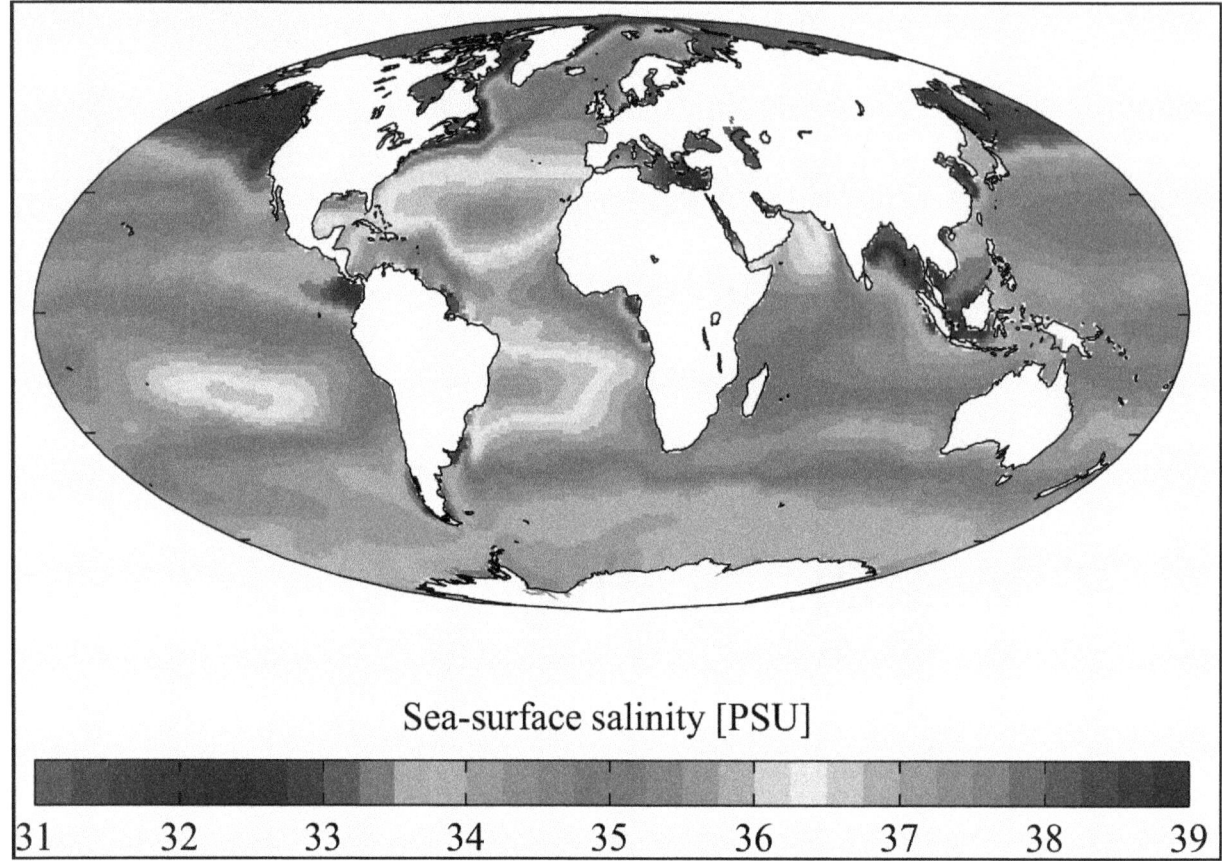

Figure 2-1   Sea-surface salinities (World Ocean Atlas 2001)

Table 2-1   Seawater mineral quality compared to national source waters

Constituent	Seawater	U.S. Freshwater Supplies	
	Average	Median	Upper Quartile
	mg/L	mg/L	mg/L
Chloride	19,350	10	20
Sodium	10,710	30	80
Sulfate	2,690	30	75
Magnesium	1,304	10	20
Calcium	419	40	60
Potassium	390	2	4
Alkalinity	146	190	250
Bromide	70	0.02	0.06
Boron	4.4	0.08	0.2
Fluoride	1	0.2	0.4
TDS	35,079	330	480

Adapted from Dittmar 1884 and James M. Montgomery 1985
Courtesy of Greg Wetterau and Shane Trussell

While the concentrations of dissolved inorganic constituents are relatively consistent throughout the open ocean, concentrations of suspended materials and colloids will vary drastically between locations and over time. As an example, the turbidity of seawater measured over one year of pilot testing at the Santa Cruz, California desalination pilot varied from 2 to 30 nephelometric turbidity units (ntu) with no significant storm events or red tide blooms experienced (Desormeaux et al. 2009). In contrast, a longer pilot study conducted by West Basin Municipal Water District near Los Angeles, California never saw turbidity exceed 10 ntu, even during a severe red tide event (Lauri et al. 2010).

Seawater turbidity can consist of suspended sediment from surface water runoff or the ocean floor, or from phytoplankton, marine larvae, and other organic matter. Similar to the variability in turbidity, the concentration of total organic carbon (TOC) varies with storm events, ocean currents, and periodic algal blooms. Turbidity, TOC, and various biological parameters can present significant challenges to seawater desalination, requiring optimized pretreatment processes or the use of subsurface intakes to provide acceptable quality water to the reverse osmosis (RO) membranes. In addition, the entrainment of marine larvae in seawater intakes has raised environmental concerns, in some cases requiring that mitigation measures be taken to address the problem. Pretreatment alternatives are discussed in detail in Chapter 3. Environmental concerns are discussed in Chapter 4.

PRODUCT WATER QUALITY

The treatment processes used in seawater desalination result in a product water that has been stripped of nearly all the mineral content, with the exception of a few constituents that are either insufficiently removed by RO membranes (such as boron and bromide) or are exceptionally high in the source water (such as sodium and chloride). These water quality characteristics result in unique water that must be further conditioned, treated, and possibly blended to produce a finished water that is acceptable for potable water systems. The key issues that must be addressed include:

1. Health concerns
2. Product water stability
3. Irrigation and industrial use concerns
4. General aesthetic concerns

Each of these issues is discussed briefly in the following section.

HEALTH CONCERNS

Issues related with public health in drinking water are regulated under the United States Environmental Protection Agency (USEPA) as administered through state health and environmental protection departments. While the requirements for seawater desalination facilities are the same for all public drinking water systems, the key health concerns with desalinated product water differ somewhat because of the unique nature of the source water and the processes used in desalination. The key health issues can be divided into (1) issues related to potentially high levels of specific minerals in the RO product, (2) issues related to pathogen removal, and (3) issues related to distribution system water quality.

Mineral Content of Product Water

Seawater RO membranes used in desalination produce high quality water that meets water quality requirements for most regulated compounds. It is generally only boron that has a significant risk of exceeding current federal or state guidelines, and these guidelines are based primarily on issues other than public health, as will be discussed later. Similar to boron, the chloride concentrations produced by seawater RO membranes may also be a concern for nonhealth related issues; however, these concentrations are unlikely to exceed state and federal guidelines for chloride. The presence of some unregulated organic compounds, such as algal toxins, has also caused some concern among State and Federal regulatory agencies, with numerous studies being done to document the removal effectiveness and prevalence of these compounds.

Boron. Because of its low molecular weight and weak ionic charge, boron is difficult to remove with most RO membranes. Typical removals vary from 40 to 90 percent at near neutral pH. Higher removals can be achieved at elevated pH, with greater than 99 percent removal reported at pH greater than 9.5 (Oo and Song 2009). With levels of boron in seawater typically at 4.4 mg/L, RO product from a single-pass treatment approach is typically on the order of 1 to 2 mg/L, depending on the temperature, flux, and specific membranes employed.

Animal tests with boron have shown adverse effects on the reproductive systems of rats and male dogs; however, the USEPA recently decided not to regulate boron with a maximum contaminant level (MCL), due to its low prevalence at levels of concern in U.S. drinking water supplies (USEPA 2007). The USEPA has established a long-term health advisory level for children of 2 mg/L and for adults at 5 mg/L, based on health effects observed during animal testing. In addition, several states (California, Florida, Maine, Minnesota, New Hampshire, and Wisconsin) have adopted standards, guidelines, or notification levels for boron ranging from 0.6 to 1 mg/L (USEPA 2008). Beyond U.S. guidelines, the World Health Organization has set a health guideline for boron since 1993. This was set initially at 0.3 mg/L, was increased to 0.5 mg/L in 1998, and is expected to increase to 2.4 mg/L in 2011.

Chloride. Chloride is generally not raised as a health issue for drinking water but is established as a secondary MCL (set at 250 mg/L), primarily for aesthetic concerns related to the taste of high chloride water. Because of the high concentration of chloride in seawater, RO product from a single-pass system can range from 100 to 200 mg/L, depending on the temperature, flux, and specific membranes employed. Lower levels may be sought by some utilities based on aesthetic concerns.

Nonregulated parameters. A large number of emerging contaminants or contaminants of emerging concern are being studied in terms of their health effects and prevalence in U.S. water supplies. While some of these contaminants can be found in seawater sources at concentration approaching those in freshwater supplies, most are considerably lower in seawater and are less of a concern to seawater desalination than they are to conventional water treatment facilities. However, compounds that are of unique concern to seawater desalination are currently unregulated toxins related to periodic algal blooms or red tide events. Two common algal toxins detected in U.S. waters are found off the Pacific Coast and include domoic acid, the cause of amnesic shellfish poisoning, and saxitoxin, the cause of paralytic shellfish poisoning. The toxins are typically not found in the water column at concentrations that are considered toxic but have been found to accumulate in shellfish to levels that are toxic to both humans and other mammals. The algae that produce the toxins are readily removed by the pretreatment filters within a desalination plant and by the RO membranes; however, there is a concern that cells will break during the pretreatment filtration process and release more dissolved toxins into the water. Pilot studies in Southern California (Carlsbad and West Basin) have demonstrated excellent removal of the toxins during large harmful algal blooms; also, a spiking of dissolved toxins at 1,000 times typical concentrations during pilot testing in Santa Cruz demonstrated greater than 3 log (>99.9%) rejection using kainic acid, a common surrogate for domoic acid.

Pathogen Removal

The USEPA Surface Water Treatment Rule (SWTR) requires removal and inactivation of viruses and *Giardia* cysts, as well as turbidity reduction, for surface waters and ground waters under the direct influence of surface water (GWUDI). The Long Term 2 Enhanced Surface Water Treatment Rule (LT2ESWTR) sets removal and inactivation requirements for *Cryptosporidium*. Seawater is currently regulated as a surface water supply in the U.S.; however, the biological activity in seawater and the survival rate of common pathogens is considerably different in seawater than in natural freshwater bodies. The survival rates of *E.Coli*, *Giardia* cysts, and *Cryptosporidium* oocysts have been shown to be low in seawater when compared with freshwater systems. These pathogens have, nonetheless, been found to be present in seawater bodies directly impacted by wastewater and stormwater outfalls (Fayer et al. 1998).

Pathogen removal requirements established by the USEPA require a multiple barrier approach to provide both removal and inactivation of pathogens. The SWTR requires a minimum of 99.99 percent (4-log) reduction in viruses and 99.9 percent (3-log) reduction in *Giardia* through filtration and disinfection. The LT2ESWTR requires between 2-log and 5.5-log reduction in *Cryptosporidium*, depending on the prevalence of *Cryptospordium* found in two years of source water monitoring, and on the type of treatment employed. In addition to the federal requirements, some states include multiple barrier requirements for specific pathogens (*Giardia* and viruses) and may require higher removal for impaired water bodies where high concentrations of total coliforms have been measured in Watershed Sanitary Survey (WSS) monitoring. Table 2-2 summarizes the minimum and maximum reduction requirements for *Giardia*, *Cryptosporidium*, and viruses.

Distribution System Water Quality

Distribution system water quality concerns relate both to corrosion within the distribution system and to the formation of disinfection by-products (DBPs). Corrosion issues, while relating to health concerns about lead and copper leaching from residential plumbing, also have aesthetic components and are discussed in the section on Product Water Stability. This section will focus primarily on the impact of desalinated water on DBP formation. The formation of DBPs, such as trihalomethanes (THMs), is impacted primarily by the presence of natural organic matter (NOM) when free chlorination is used for disinfection. Levels of NOM in desalinated product water are considerably lower than in most conventional treatment products. The formation of DBPs is therefore a concern primarily when desalinated seawater is blended in a distribution system with treated surface waters containing significant levels of NOM.

Table 2-2 Pathogen reduction requirements for surface waters

Pathogen	Log Reduction Requirement		Criteria for Determination
	Minimum	Maximum	
Virus	4	6	Total coliforms[1]
Giardia	3	5	Total coliforms[1]
Cryptosporidium	2	5.5	*Cryptosporidium*[2], total coliforms[1]

1. California Department of Public Health requires up to 2-log additional removal based on high levels of total coliforms.
2. LT2ESWTR requires up to 5.5 log *Cryptosporidium* removal, based on bin classification from *Cryptosporidium* monitoring

Courtesy of Greg Wetterau

20 DESALINATION OF SEAWATER

Source: HortScience Inc.

Figure 2-2 Boston Ivy with tip burn from chloride

Source: HortScience Inc.

Figure 2-3 Boron toxicity on camphor (cinnamomumcamphora)

As noted previously in Table 2-1, seawater contains unusually high levels of bromide, when compared with most freshwater sources in the U.S., with the concentration of bromide roughly three orders of magnitude larger in seawater than in the highest quartile of freshwater supplies. Bromide removal with seawater RO membranes is comparable to the removal of chlorides; however, the level of bromide that results from seawater treatment in a single-pass system is still an order of magnitude larger than in most freshwater supplies (ranging from 0.5 to 0.8 mg/L under typical operating conditions). High bromide levels have been associated with the formation of brominated DBPs. Chlorine added for disinfection can oxidize bromide to bromine; the bromine may then react with NOM to form brominated DBPs. Brominated DBPs are a concern because of an increased health risk compared with chlorinated DBPs, and because brominated compounds increase the mass-based concentration more than chlorinated DBPs due to the molecular weight of bromide, which is approximately double that of chloride.

Brominated DBPs can be expected to form in zones where desalted water blends with other supplies when free chlorine is used for disinfection residual maintenance. DBP formation tests are commonly recommended with blends of the desalinated water and local treated surface water to assess if this is a potential issue. Previous studies of desalinated water disinfection have also shown that high levels of bromide can have adverse impacts on the stability of chloramine residual in distribution systems (USEPA 2001).

PRODUCT WATER STABILITY

Seawater desalination results in a product water quality that is characterized by low hardness (calcium and magnesium), low alkalinity, and relatively high chloride. The hardness and alkalinity are so low that desalination facilities typically condition the product water to achieve a slightly positive Langelier Saturation Index (LSI) to control corrosion in the distribution system and consumer plumbing. Soft water supplies are well known to have a higher potential to create corrosion-related apparent color (red water) problems when conveyed in iron pipe and to cause leaching of lead and copper in home plumbing, which can increase health risks and violate requirements of the Lead and Copper Rule. High chloride content may also exacerbate these problems, increasing the corrosivity of the water. Cast-iron or ductile-iron pipes, either unlined or lined with older epoxy coatings, are susceptible to iron release and red water complaints if the softened water is not appropriately conditioned for corrosion protection.

The LSI, an index first defined by Wilfred Langelier in 1936 (Langelier 1936), has become the most widely used measure of product water aggressivity, and a positive LSI is used as a target for many desalination post-treatment facilities. Other measures of product water aggressivity include the calcium carbonate precipitation potential (CCPP), which is often kept between 4 and 10 mg/L, and the aggressiveness index (AI), which is often kept above 12. The adjustment of $CaCO_3$ saturation, however, is not uniformly practiced nor is it always successful in meeting corrosion control objectives. Other common approaches for mitigation of corrosive effects from soft water can include pH adjustment and addition of orthophosphate. It is generally most important, however, to avoid significant variations in either pH or calcium carbonate saturation to prevent problems with pipeline corrosion or customer complaints from cloudy or red water.

IRRIGATION AND INDUSTRIAL USE CONCERNS

While desalinated product water from seawater supplies has a low risk of exceeding health related water quality goals, the levels of boron, chloride, and sodium in the product of single pass RO system could create significant problems for irrigation and industrial users. The primary issues related to these types of users are discussed in the following sections.

Irrigation and Agricultural Use

High levels of chloride, sodium, and boron are known to impact numerous different ornamental plants, citrus trees, and other common agricultural crops. Although the total dissolved solids (TDS) concentration is low in desalinated ocean water, the concentrations of chloride and sodium comprise a majority of the TDS. Chloride and sodium toxicity have been more commonly observed than boron toxicity to date because of the increased use of reclaimed water for irrigation purposes, particularly in the arid Southwest. Some of the better-known cases of chloride toxicity have occurred where irrigation of redwood and avocado trees was converted from local water supplies to recycled water, resulting in significant impacts on the health of these species. The most common symptom of chloride, sodium, and boron toxicity is tip burn, but chloride or sodium toxicity can also result in tattered leaves, reduced leaf size, reduced growth rate, yellowing of leaves in conifers, and in extreme cases, plant death. Similar to boron, chloride and sodium accumulate in the older leaves, and these leaves typically exhibit the symptoms first. Figure 2-2 shows an example of chloride toxicity observed in Southern California and caused by irrigation using reclaimed water with high chloride content.

Concerns with boron may arise if there is a significant difference between the boron content of the traditional water supply source or sources and the desalinated water, and if chloride and boron-sensitive plants and crops are present in the service area. Figure 2-3 shows mild and advanced stages of boron toxicity.

Boron concentrations greater than 2 mg/L are unhealthy for most plants, while a boron concentration greater than 1 mg/L may still have an impact on sensitive plant species. If the boron concentration of water used for irrigation is maintained below 0.5 mg/L, boron toxicity should not result in significant impacts on plant appearance. Water with boron concentration between 0.75 and 1.0 mg/L may still impact some trees, plants, and ornamentals that are commonly used for landscapes and residences in southern coastal areas of the U.S. where seawater desalination is being considered. Some of these plants include gardenia, crape myrtle, camellia, giant birds of paradise, heavenly bamboo, hydrangea, lemons, lily of the Nile, oranges, philodendron, photinia, pink trumpet vine, southern magnolia, violet trumpet vine, dwarf pittosporum, xylosma, and various roses (Yermiyahu et al. 2005).

Yermiyahu et al. (2007) reported that the seawater desalination plant (73 mgd) in Ashkelon, Israel is producing desalinated water with a boron level of 0.2 to 0.3 mg/L to meet Israeli recommendations for domestic and agricultural usage. It should be noted that a large portion of the water produced by the Ashkelon plant is used for irrigation of citrus crops, which are sensitive to boron levels in the water of more than 0.75 mg/L. The main rationale for producing a water with half of this boron level was to reduce the overall boron level of existing surface water supplies of naturally high boron content with which this desalinated water is blended. For similar reasons, Tampa Bay Water has targeted the chloride level in their desalinated water to be less than 100 mg/L. This high-quality water is blended with other surface water and groundwater sources, which have significantly higher chloride and TDS levels, improving overall water quality in the distribution system in terms of chloride, boron, and TDS.

Industrial Use

The level of chloride in the treated water may also be an important consideration for industries located within the service area of the desalination plant. As previously mentioned, the maximum chloride concentration in desalinated seawater treated using only a single-pass RO system is significantly higher than the concentrations observed in conventional water supplies (typically 150 to 220 mg/L v. less than 100 mg/L). It is known that some industries (i.e., petrochemical, power generation, pharmaceutical, breweries, and semi-conductor wafer manufacturing) use high purity water, which has chloride levels significantly lower than 100 mg/L. If these types of industries are large water users in the desalination plant service area, product water quality targets in terms of chloride, TDS, and other compounds may need to be adjusted to accommodate their requirements. An example is Contra Costa Water District in Northern California which has set a water quality objective of 100 mg/L chloride based on the requirements of the industries served in the desalination plant service area. The Water and Sewarage Authority of Trinidad and Tobago has set a chloride objective of 80 mg/L for their 36 mgd Point Lisas seawater desalination project based on the requirements of the industries served there. The Point Lisas plant is located in an industrial park where over 80 percent of its water is used by local industries.

It should be noted that many of the industrial plants that have stringent requirements in terms of TDS, chloride, or other compounds are often equipped with water purification facilities, and it is often much more cost effective for these installations to be upgraded or modified than to impose a significant incremental cost increase for a more elaborate desalination system to the general population. This is especially true when industries with very stringent product water quality requirements constitute a relatively small portion of the water demand in the plant service area.

Addressing Irrigation and Industrial Use Concerns

To address the issues with boron, sodium, and chloride as previously discussed, several seawater desalination plants worldwide have been designed with RO systems that have a partial or full second pass, where a portion of the permeate produced by the first-pass RO system undergoes additional RO treatment. By using a partial or full second pass, boron and chloride levels in the finished product water can be targeted to specific needs of irrigation and industrial users in the area served by the desalination facility.

Examples of facilities that have a partial or full RO second pass include Tampa, Florida (25 million gallons per day (mgd) or 95 megaliters per day (MLD)–50 percent second RO pass); Trinidad and Tobago (36 mgd or 136 MLD–100 percent second RO pass); Fukuoka, Japan (13 mgd or 49 MLD–100 percent second RO pass); and Tuas, Singapore (36 mgd or 136 MLD – 100 percent second RO pass). In contrast, most of the seawater desalination plants in Spain, Cyprus, and Malta, providing water for both human consumption and agricultural irrigation for grain crops and vegetables, have single-pass SWRO systems only. Most municipal SWRO treatment plants in the Caribbean are also built with a single-pass SWRO system.

GENERAL AESTHETIC CONCERNS

Although aesthetics are typically considered better for desalinated product water than most conventional water supplies, some concern should be considered when variability is possible in the temperature or taste and odor of the water delivered to residential customers. These two issues are discussed briefly in the following sections.

Temperature

Ambient ocean water temperature is typically lower than or equal to that of most conventional surface water sources. Membrane desalination processes do not result in significant increases in product water temperature, and therefore, desalinated seawater is typically of temperature comparable to that of other water sources. However, in the case of seawater desalination plants that are collocated with power plants, where cooling water from a power plant is used as a source for desalination, the product water may be 2 to 5°C higher than the ambient ocean water temperature. Therefore, in this case, the temperature of the desalinated water may be higher than that of the existing water sources, which may have an impact on water quality aesthetics. The temperature aspect of desalinated seawater should be evaluated on a case-by-case basis. For example, experience at the Tampa Bay Water seawater desalinated plant and environmental impact reports of the collocated Carlsbad and Huntington Beach seawater desalination plants indicate that the desalinated water temperature is well within the temperature of all other water sources in the desalination plants service area.

Taste and Odor

Depending on how the desalinated water is introduced to the distribution system, water supply to some consumers may be converted entirely from the conventional water source to the new desalinated water. If these consumers remain on the new supply 100 percent of the time, following a transition period, the number of taste and odor complaints would not be expected to be significant as a result of the conversion to desalinated water. It is common, however, that on days of peak water demand or days when portions of a desalination plant are down for maintenance, a water utility may need to supplement their desalinated water supply with conventional water supplies, creating a risk of taste and odor complaints. It is important to develop an appropriately designed connection point and distribution system infrastructure that prevents consumers from transitioning between the two supplies on a routine basis. The number of taste and odor complaints would be expected to increase proportionally with the frequency of that transition. This means that each time conventional water supplies are required to supplement the desalinated water, there would be two sensitive periods where taste and odor complaints would be likely to occur. Any significant transition in water supply will be noted by consumers, and taste and odor complaints will become a significant issue if the delivered water source is frequently changed. This issue can be addressed by ensuring adequate blending and mixing with a local supply prior to distribution, which may allow for providing a consistent water supply. Such an approach was taken for the Sand City coastal desalination facility near Monterey, Calif., where up to 50 percent distribution system water is blended with the desalination product to provide a consistent water quality within the system.

REFERENCES

American Water Works Association, 2008. *Manual of Water Supply Practices M46 Reverse Osmosis and Nanofiltration*. Denver, Colo.: American Water Works Association.

Antonov, J. I., R. A. Locarnini, T. P. Boyer, A. V. Mishonov, and H. E. Garcia, 2006. *World Ocean Atlas 2005, Volume 2: Salinity*. S. Levitus, Ed. NOAA Atlas NESDIS 62, U.S. Government Printing Office, Washington, D.C

Crittenden, J.C., R.T. Trussell, D.W. Hand, K.J. Howe, and G. Tchobanoglous, 2005). *Water Treatment: Principles and Design*. New York: John Wiley & Sons, Inc.

Desormeaux, E., P. Meyerhofer, and H. Luckenbach. 2009. Comparing Biological Filtration and Ultrafiltration as Pretreatment for Seawater Desalination. Proc. of AWWA Membrane Technology Conference, March 2009.

Dittmar, W. 1884. "Report on researches into the composition of ocean water collected by HMS Challenger, during the years 1873-1876." *Report on the Scientific Results of the Voyage of HMS, Physics and Chemistry*, 1(1) 251

Fayer, R., T.K. Graczyk, E.J. Lewis, J.M. Trout, and C.A. Parley. 1998. "Survival of infectious Cryptosporidium parvum oocysts in seawater and Eastern oysters (Crassostrea virginica) in the Chesapeake Bay." *Applied and Environmental Microbiology*, 64(3), 1070-1074.

James Montgomery, Consulting Engineers, Inc. 1985. *Water Treatment Principles and Design*. New York: John Wiley & Sons, Inc.

Langelier. W.F. 1936. "The analytical control of anti-corrosion water treatment." *Jour. AWWA*, 28(10), 1500-1521.

Lauri, P., M. Donovan, J. Dietrich, and S. Trussell (2010) "Comprehensive Results of West Basin Municipal Water District's Ocean Water Desalination Pilot Program," Proceedings Annual WateReuse Symposium, Seattle, WA.

Matheny, N., Evaluation of Proposed Irrigation Water Quality on Carlsbad Landscapes. 2005, Poseidon Resources Corporation: Stamford, CT.

Miller, J.E., and T.M. Mayer. 2005. "A Critical Review of Alternative Desalination Technologies." State-of-the-Science Report Number 2, Joint Water Reuse Desalination Task Force.

Oliver, J.D. 2005. "The viable but nonculturable state and cellular resuscitation." International Symposium on Microbial Ecology, Halifax, Nova Scotia, Canada.

Oo, M.H. and L. Song. 2009. "Effect of pH and ionic strength on boron removal by RO membranes." *Desalination*, 246(1-3), 605-612.

Prato, T., E. Schoepke, L. Etchinson, T. O'Brien, B. Hernon, K. Perry, and M. Peterson. 2001. "Successful use of membranes at Diablo Canyon nuclear plant." *International Desalination & Water Reuse Quarterly*, 10(4), 27-31.

Stumm, W. 1960. "Investigation of the corrosive behaviour of waters." *Jour. Sanitary Eng.* Division ASCE, 86(11), 27-45.

Stumm, W. 1956. "Calcium Carbonate Deposition at Iron Surfaces." *Jour. AWWA*, 48(3), 300-310.

Thompson, J.D., J. Arnold, and I. Moch. 2005. "Three years operation of the 27.6 MIGD, desalination plant, Point Lisas, Trinidad and Tobago." IDA World Congress on Desalination and Water Reuse, Singapore.

Trussell, R.R., J. Jacangelo, and R. Cass. 2005. "Design and performance of the pretreatment for the Point Lisas desalter." AWWA Conference, Orlando, Fla.

USEPA. 2002. Mechanisms and Kinetics of Chloramine Loss and By-Product Formation in the Presence of Reactive Drinking Water Distribution System Constituents. http://cfpub.epa.gov/ncer_abstracts/index.cfm/fuseaction/display.abstractDetail/abstract/205/report/F

USEPA. 2005. Membrane Filtration Guidance Manual. Washington, D.C.: USEPA.

USEPA. 2007. Preliminary Regulatory Determinations for Priority Contaminants on the second Contaminant Candidate List, Federal Register May 7, 2007. Washington, D.C.: USEPA.

USEPA. 2008. Drinking Water Health Advisory for Boron, U.S. Environmental Protection Agency, Office of Water, Document No. 822-R-08-013, May. Washington, D.C.: USEPA.

Yermiyahu, U., A. Ben-Gal, and P. Sarig. 2006. Boron Toxicity in Grapevine. *HortScience* 41: 1698-1703.

Yermiyahu, U., A. Tal, A. Ben-Gal, A. Bar-Tal, J. Tarchitzky, and O. Lahav. 2007. "Environmental science: Rethinking desalinated water quality and agriculture." *Science*, 318, 920-921.

AWWA MANUAL M61

Chapter 3

Treatment Approaches

Larry VandeVenter
Robert Huehmer
Val Frenkel

PRETREATMENT

Historically, effective pretreatment has been the most challenging issue confronting users of seawater reverse osmosis (SWRO) systems. Pretreatment of seawater is often necessary prior to its application to RO membranes to remove potential foulants such as particulate, colloidal, and biological material that can potentially scale/foul the membranes. Membrane fouling can lead to increased feed pressure requirements, increased energy consumption, reduced permeate production, reduced permeate quality, and membrane damage. Pretreatment requirements for seawater desalination facilities are determined by the type of seawater intake and influent water quality. The type of seawater intake influences the influent quality which in turn affects the choice of pretreatment processes, which can include chemical addition, clarification, media filtration, and/or membrane filtration. Chemical addition in the pretreatment process may include coagulants and filtration aids as well as acid and/or scale inhibitor, which are added to reduce scaling of sparingly soluble salts on the RO membrane surface. Disinfection, either continuous or intermittent shock, is often included in the pretreatment process as well to prevent biological fouling, and dechlorination ahead of the RO membranes is required to prevent damaging the RO membranes. This section presents common approaches to seawater reverse osmosis (SWRO) pretreatment, RO design parameters, disinfection, posttreatment, energy recovery, and corrosion and materials of construction.

Water Quality of Pretreated Water

SWRO desalination systems use semi-permeable membranes to produce high quality permeate. Because RO membranes generally reject high percentages of salts (> 99 percent), salts accumulate on the feed side of the membrane and the solubility of sparingly soluble salts such as calcium sulfate, barium sulfate, strontium sulfate, or calcium carbonate may occur, resulting in precipitation of these salts on the membrane surface. Precipitation of these salts increases the feed pressure requirements to move water through the membranes, which results in higher operating costs, increased salt concentrations in the permeate stream, increased cleaning requirements, and potential membrane damage.

SWRO system operators can control the concentration of sparingly soluble salts in the concentrate stream by controlling the system recovery, where a reduction in recovery lowers their concentration in the concentrate, thereby lowering the potential for membrane scaling. Another approach to minimize scaling is chemical conditioning of the feed water using scale inhibitors and dispersants; these chemicals can prevent or delay precipitation under certain levels of supersaturation for some sparingly soluble species. Natural organic matter in seawater is believed to provide some level of natural scale inhibition (Chave and Suess 1970). As a result, many seawater installations have operated with no scale inhibitors at recoveries on the order of 50 percent with sparingly soluble salt concentrations exceeding equilibrium conditions by as much as 300 percent. The ultimate recovery of SWRO units is typically determined by operating pressure and salt passage. Detailed computer simulations can be performed to project system recovery based on water quality and operating conditions.

The primary indicator of SWRO pretreatment effectiveness is the Silt Density Index (SDI). This simple analytical technique provides a qualitative measure of the fouling potential of the tested water. RO membrane manufacturers typically recommend SDI values less than 4.0 while a value of less than 3.0 is preferable. The SDI is not a perfect indicator of potential fouling, but it is the industry standard for indirect measurement of particulate and silt content in the source seawater currently used. In general, lower SDI measurements indicate lower fouling potential that can result in longer intervals between membrane cleanings. Subjecting SWRO membranes to the rigors of chemical cleaning generally lowers the membrane life expectancy and increases the overall O&M costs of the system.

Pretreatment Technologies

The first commercial SWRO plants were installed in Saudi Arabia in 1975, and there are currently more than 1,000 seawater RO plants constructed worldwide. The pretreatment utilized varies significantly according to the type of seawater intake method. There are three general types of seawater intakes: (1) open ocean intakes, (2) beach wells, and (3) infiltration galleries. A detailed discussion of intake alternatives is included in Chapter 4 of this manual. Generally, seawater drawn directly from open intakes requires more robust pretreatment than seawater derived from beach wells or infiltration systems, so pretreatment for SWRO applications can be divided into two categories in relation to the feed water supply system: (1) surface supply intakes, and (2) subterranean or subsurface intakes. The pretreatment requirements vary greatly as a function of which supply system is utilized. The following sections describe pretreatment approaches for both categories of intake facilities.

Pretreatment for subterranean intake facilities. Pretreatment in beach well applications is typically limited to chemical addition for scale inhibition and cartridge filtration. Limited pretreatment is generally required as a result of low levels of particulate, colloidal, and biological material after the seawater has been prefiltered through the sand of the seafloor or beach. Additionally, naturally filtered seawater generally has relatively constant physical characteristics such as temperature, turbidity, and microbial content. However, in cases when beach wells (one type of subterranean intake) collect water from alluvial coastal aquifers, source waters may contain iron and manganese in quantities that could require pretreatment with green sand filters or other oxidation/filtration processes.

Little information is available on the reliability of water quality from infiltration galleries, because of their limited use in full-scale desalination facilities. Preliminary testing at a pilot facility in Long Beach, Calif. has suggested that an infiltration gallery with a minimum 5-feet (1.5 meters) of cover could provide a source water with a turbidity less than 1.0 nephelometric turbidity units (ntu) (Cheng et al. 2010).

Pretreatment in surface supply intake facilities. Where open seawater intakes are used, potential worst-case water quality issues that could substantially affect the ultimate plant cost are listed below. These parameters should be considered as indications that increased pretreatment is required to ensure stable SWRO operation:

- turbidity greater than 20 ntu
- measurable levels of hydrocarbon-based contaminants
- significant occurrences of red tides or algae
- high levels of pathogens
- large variations in temperature of the raw water
- moderate levels of total organic carbon (TOC)
- severe water quality excursions caused by hurricanes or other severe storm events

Pretreatment system processes for open-intake-derived seawater must be robust enough to handle expected worst case variations in water quality and provide low levels of SDI for stable operation of the RO process.

Conventional pretreatment unit operations for open intake seawater SWRO plants have consisted of the following processes, each of which is described in Table 3-1.

- Chlorination
- Coagulation, flocculation, and clarification
- Filtration
- Chemical dosage for scale inhibition
- Cartridge filtration
- Dechlorination

In certain limited cases, additional pretreatment processes have been implemented, such as diatomaceous earth filtration and granular activated carbon (GAC). GAC is used most often in Arabian Gulf applications to remove oil and grease that may be present in the feedwater. In addition, several novel approaches to SWRO pretreatment are in the development or demonstration stages. Some of these approaches are listed below and are discussed in Table 3-2.

- Upflow solids contact clarification – Pulsator or SuperPulsator
- Dissolved air flotation (DAF)
- Membrane ultrafiltration (UF) and microfiltration (MF)
- Micro-sand enhanced clarification (MES) – Actiflo

A further description of pretreatment processes used in desalination can be found in the AWWA Manual of Practice on Reverse Osmosis and Nanofiltration (M46).

Pretreatment at recently installed SWRO plants. Outside the US, SWRO has typically employed direct filtration as part of the pretreatment process. Single or two-stage direct filtration can only be used when the raw water turbidity is relatively low. Water with higher levels of turbidity or algae may require clarification prior to filtration. For example, a large SWRO plant in Trinidad is operating with tube settler clarification and single-stage filtration; this pretreatment regime was considered necessary due to raw water turbidity excursions as high as 35 ntu where the typical turbidity is generally less than 10 ntu. The design loading rate of the tube settlers is 1.9 gallons per minute per square foot (gpm/f^2) or 4.6 meters per second (m/s) and the filters are designed at 6.5 gpm/f^2 (15.9 m/s).

Table 3-1   Seawater RO pretreatment components for surface seawater sources

Pretreatment Component	Description	Discussion
Chlorination	Chlorination is frequently used for disinfecting the intake and pretreatment system to mitigate biofouling in the downstream RO. Historically, continuous chlorination was used at levels up to 5 mg/L, where cellulose acetate membranes were employed. For chlorine sensitive thin-film-composite membranes, the use of intermittent shock chlorination has become more common for controlling biofilm growth.	It was believed that continuous chlorination was necessary to prevent RO biofouling. Chlorination of naturally occurring humic and fulvic acids create high concentrations of assimilable organic carbon (AOCs), which is currently known to be a principal player in the RO biofouling process. Intermittent shock chlorination has shown to be an improvement in many plants, while some have totally eliminated disinfection with successful results.
Coagulation/ Flocculation/ Sedimentation	Coagulation and flocculation are used to destabilize the suspended and colloidal material from the raw seawater. The most common coagulants include ferric salts such as ferric chloride and ferric sulfate (typically used at levels of 3-10 mg/L as Fe). Multiple flocculation stages followed by sedimentation have been used successfully. Inline coagulation is commonly used in treating water with low fouling potential.	Highly variable water quality may require coagulation/flocculation/ sedimentation unit processes. Sufficient mixing is critical, especially when only inline coagulation is followed by direct filtration. Conventional sedimentation operates at 0.5 gpm/f^2, and has a large footprint. Tube and plate settlers have been used at loading rates of 1.0 – 3.0 gpm/f^2, and 3.0 – 5.0 gpm/f^2, respectively.
Filtration	Media filtration typically involves dual media sand and anthracite, or mixed media sand, anthracite, and garnet as media. Both single and two-stage systems are common in SWRO, as are both pressure and gravity filters. Typical loading rates are 2-6 gpm/ft^2. Nonionic and anionic polymers are sometimes used as a filter aids.	Plants that use inline coagulation may use two-stage filtration to address the higher solids loading from nonclarified water. SDI goals of less than 3.0 units are generally achievable with sufficient design in the coagulation and filtration processes.
Chemical Addition	Sulfuric acid is often used to depress the pH to prevent calcium carbonate scaling. Scale inhibitors have also been applied to prevent or delay sparingly soluble sulfate salts from precipitating.	Acid addition has not been shown to be problematic but has been replaced more recently by scale inhibitors.
Cartridge Filtration	Cartridge filters are used as the last line of defense against particles reaching the RO membrane surface. Typically, 5 micron filters are used; occasionally 1–3 micron are used, especially during initial commissioning.	When the coagulation/filtration processes have not been sufficiently designed, the cartridge filters incur high loadings requiring frequent replacement. Iron deposits and biofouling are frequent complaints in a poor performing plant.
Dechlorination	RO membranes are susceptible to chlorine oxidation, and therefore all chlorine must be removed from the pretreated water. Sodium bisulfite (SBS) is the most commonly used dechlorinating agent, with doses of 3–4 mg/L typically used.	Rapid biofouling occurs immediately following dechlorination in plants with continuous chlorination and high organic content. Additionally, reduction of ferric salts can create catalyzed chlorine oxidation. SBS alone is not problematic and has shown to have biostatic properties.

Courtesy of Larry VandeVenter

Table 3-2 Seawater RO treatment advancements for surface seawater sources

Pretreatment Component	Description	Discussion
Up-Flow Solids Contact Clarification	Up-flow solids contact clarifiers combine coagulation and flocculation within a pulsed blanket and are widely utilized in fresh water treatment. They combine coagulation and flocculation within a pulsed blanket.	Up-flow solids contact clarifiers operate at 1-2 gpm/f^2 and have a smaller footprint than conventional flocculation and sedimentation; however, these have not been commonly used in desalination applications.
Dissolved Air Flotation (DAF)	DAF is used for clarification upstream of conventional filtration as a stand-alone clarification process or can be stacked over filtration. DAF uses micro bubbles to float coagulated solids, colloidal material, and algae. DAF can be operated without a polymer.	DAF operates at 6-12 gpm/f^2, significantly higher loading rates than conventional sedimentation or plate settlers and has a smaller footprint. Pilot data has shown DAF to improve pretreated water quality especially in removal of algal content (red tide). DAF has been successfully employed with SWRO at the Tuas facility in Singapore.
Membrane Filtration	Membrane filtration uses microfiltration or ultrafiltration to replace the flocculation/sedimentation and filtration processes of conventional pretreatment; or to replace just the filtration process following some form of coagulation or clarification.	Multiple vendors of MF and UF are available in a variety of formats–either pressure or immersed. Pilot studies have shown reduced or eliminated coagulant dosage using membrane filtration and enhanced pretreated water quality, with SDI values generally around 2. Membrane filtration has been successfully employed with SWRO in Australia and the Mediterranean.
Micro-sand enhanced settling (MES)	MES is used for clarification upstream of conventional filtration. MES uses micro-sand attached to the coagulated solids by a polymer to enhance the clarification process.	MES operates at 16-35 gpm/f^2, loading rates significantly higher than conventional sedimentation, plate settlers, and DAF, and has a very small footprint. MES has not yet been employed at full-scale SWRO facilities.

Courtesy of Larry VandeVenter

In the U.S., the largest SWRO facility (Tampa Bay, Fla.; 25.0 mgd/95 MLD) initially used direct filtration with two-stage continuously backwashing upflow filters. However, during preliminary operation of this plant, there were severe problems with the pretreatment process because of poor feedwater quality, including growth of mussels and problems with mussel shell fragments. Corrective measures included conversion of the original two-stage sand filtration system into single-stage filters at a reduced filter loading rate, followed by diatomaceous earth filtration as an additional barrier to prevent mussel related issues.

Recently installed SWRO plants have used a variety of different pretreatment approaches depending on the type of seawater intake and the feedwater quality. Table 3-3 summarizes some of these installations (IDA Desalination Yearbook 2009-2010). Pretreatment systems have included single and two-stage filtration, sedimentation, dissolved air flotation, ultrafiltration, and diatomaceous earth filtration. Both gravity and pressure filtration have been installed.

In facilities employing ultrafiltration or microfiltration, an additional screening step is required to prevent damage to the membranes. While 500 micron strainers are commonly used in freshwater treatment facilities employing MF and UF, the prevalence of barnacles in seawater, which can be 130-150 micron during their embryonic phase, necessitates the use of tighter micro-strainers with mesh sizes between 80 and 120 micron. Micro-strainers which have been employed or successfully piloted at SWRO facilities include plastic rotating disc strainers, rotating drum screens, and automatic backwashing strainers.

Table 3-3   Partial list of pretreatment installations in SWRO plants since 1995

Installed Date	Location	Plant Name	Capacity MLD	Capacity mgd	Intake Type	Pretreatment
2009	Australia	Gold Coast	133	35.1	Offshore, open sea	Single-stage dual media gravity filters
2009	UAE	Fujairah II	136	35.9	Offshore, open sea	Flocculation, DAF, single-stage dual media gravity filters
2008	Algeria	Skikda	100	26.4	Open sea	Two-stage, dual media pressure filters
2008	UAE	Sharjah	22.7	6.0	Offshore, open sea	Two-stage, dual media pressure filters
2008	Algeria	Hamma	200	52.8	Offshore, open sea	Lamella clarifiers, single-stage dual media gravity filters
2008	Aruba	Balashi	8	2.1	Shoreline, open sea	Pressurized, single stage sand filters
2007	Florida	Tampa Bay	109	28.7	Shoreline, open sea	Flocculation, single-stage upflow sand filters, DE filters
2007	Israel	Palmachim	110	29.0	Offshore, open sea	Single-stage, dual media gravity filters
2007	Australia	Kwinana (Perth)	144	37.9	Offshore, open sea	Single stage, dual media pressure filters
2007	Singapore	Power Seraya	10	2.6	Shoreline, open sea	Two-stage, dual media pressure filters
2006	Bahamas	Blue Hills	27	7.2	Seawater wells	Cartridge filtration
2006	Saudi Arabia	Kindasa	25.5	6.7	Shoreline, open sea	Single stage, dual media filters and ultrafiltration
2006	China	Yu-Han	70	18.5	Open sea	Ultrafiltration
2005	Israel	Ashkelon	326	86.1	Offshore, open sea	Single-stage dual media gravity filters
2005	Japan	Fukuoka	50	13.2	Offshore infiltration gallery	Ultrafiltration
2005	Singapore	Tuas	136	36.0	Offshore, open sea	DAF, single-stage dual media filtration
2003	UAE	Fujairah I	170	44.9	Offshore, open sea	Single-stage, dual media gravity filters
2002	Trinidad	Point Lisas	119	31.4	Shoreline, open sea	Flocculation, sedimentation, single-stage dual-media deep bed gravity filters
2002	Spain	Carboneras	120	31.7	Offshore, open sea	Single-stage, dual media pressure filters
2001	Cyprus	Larnaca	54	14.3	Offshore, open sea	Single-stage, dual media gravity filters
2000	Bahrain	Addur	140	37	Open sea	Ultrafiltration
1995	Japan	Okinawa	40	10.6	Offshore, open sea	Two-stage, pressure filters

Courtesy of Tom Pankratz/Water Desalination Report

SWRO DESIGN PARAMETERS

SWRO systems typically consist of feedwater pumps, RO membrane elements installed in pressure vessels, supporting framework and racks for the pressure vessels, valves and piping, instrumentation and controls, and sample panels. Either single or two-stage RO can be used depending on finished water quality goals. Energy recovery devices are typically used in seawater applications to reduce energy use and therefore lower operational costs. The main parameters for SWRO design include flux, recovery, and operating temperature. Each of these is discussed briefly in the following sections. In addition to these parameters, source water TDS will have a considerable impact on operating pressures and facility cost. Where source waters are available with salinity lower than the open ocean, these waters are often utilized.

RO Flux

The membrane flux normalizes the filtration rate per membrane surface area and can be expressed as gallons per day per square foot of membrane area, or gallons per square foot per day (gfd) or in metric units as liters per meter squared per hour (LMH). Flux has a major impact on both the capital and operating costs of the facility, where a lower flux will require more membrane area, higher capital cost, and larger plant footprint in comparison to higher fluxes. However, higher fluxes require higher feed pressures that result in increased energy use and higher operating costs. Use of higher fluxes can also increase fouling, causing further increase to the operating costs. The Affordable Desalination Collaboration recently investigated operations in the range of 6 to 9 gfd (10 to 15 LMH) and concluded that this range may result in lower overall life cycle costs for desalination. Most existing SWRO facilities operate at fluxes of approximately 8 to 12 gfd (14 to 20 LMH).

An additional result of operating at higher fluxes is increased salt rejection. This phenomenon may seem counter-intuitive, but it is the result of a higher ratio of product water diluting the relatively constant passage of salt across the membranes. Increasing the flux will increase this dilution factor, improving the quality of the product water. Ultimate determination of the most appropriate flux for a specific SWRO facility involves balancing capital costs, operating costs, and desired product water quality.

Recovery

Product water recovery is the ratio of the product water flow rate to the feedwater flow rate, representing the overall production efficiency of the desalination system. Higher recoveries result in smaller intakes, smaller pretreatment facilities, and lower waste flows, but also produce more concentrated residual streams and higher salinity product water. For example, 50 percent recovery will concentrate the TDS levels in the source water by twofold while 75 percent recovery will increase TDS levels fourfold. Elevated concentrations of sparingly soluble salts in the concentrate cause higher scaling potential and increased osmotic pressure requiring higher feed pressures.

Brackish water and reclaimed water treatment facilities typically operate at recoveries between 70 and 80 percent because these water sources have lower TDS levels than seawater; recovery in these cases is generally limited by the scaling potential of sparingly soluble salts that may be present. Higher TDS levels in seawater result in higher feed pressure requirements, and recoveries in SWRO facilities typically range from 40 percent to 55 percent, which in part reflects the balance between operational costs (feed pump energy demands and pretreatment requirements) and capital costs (overall plant footprint and membrane system requirements). In addition, the residuals volume (membrane concentrate) will decrease as recovery increases, so the costs of residuals management must also be considered.

Figure 3-1 provides the power consumption SWRO systems as a function of system recovery for three different feedwater temperatures (adapted from Wilf and Klinko 2001). While power consumption will vary with membrane type, flux, and energy recovery measures employed, the general trends shown in Figure 3-1 should apply.

In addition to the impact on energy use, a higher recovery will increase the salt concentrations in the permeate, resulting in a poorer product water quality.

Feedwater Temperature

For SWRO systems, an increase in feedwater temperature results in lower feed pressure requirements, and therefore lower energy use, when flux and recovery are maintained at constant conditions. Figure 3-2 shows this effect using manufacturer software projections of the feed pressure as a function of temperature for two different flux rates and two different SWRO membrane element types (Hydranautics SWC5 and SWC4+ elements, 50 percent system recovery, and typical seawater quality provided in Table 2-1). As this figure shows, an increase from 15°C to 25°C can decrease the projected feed pressure requirements by as much as 100 psi (690 kPa).

In addition to affecting feed pressure requirements, changes in the influent water temperature also affect salt passage. Higher temperatures increase the salt passage, resulting in a lower water quality (higher salt content). If boron and chloride levels are constraining parameters in product water quality, this increased salt passage can result in a water quality that may not meet the treatment goals, depending on the flux, recovery, and membrane type used.

The relationship between feedwater temperature and permeate boron concentration is presented in Figure 3-3; this figure was generated using manufacturer software projections for two membrane types and two flux rates (50 percent recovery with typical seawater quality identified in Chapter 2). This figure indicates that a boron goal of 1.0 mg/L can be achieved at temperatures below 22°C for either membrane type at fluxes greater than 10 gfd (17 LMH). Using higher boron rejection membranes, this goal can be met at temperatures as high as 34°C.

A lower boron goal of 0.5 mg/L can also be met using the high boron rejection membranes, but only at temperatures up to 15°C for a flux rate of 10 gfd (17 LMH) and for temperatures up to 20°C for a flux rate of 12 gfd (20 LMH).

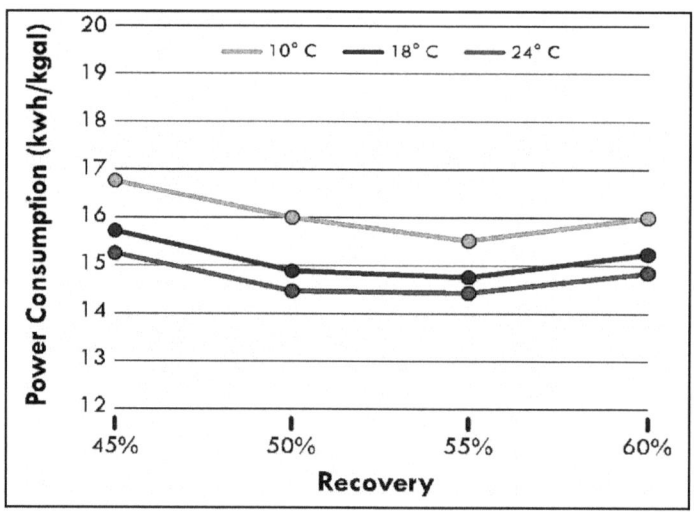

Adapted from Wilf and Klinko 2001.

Figure 3-1   Projected impact of recovery on power consumption for SWRO

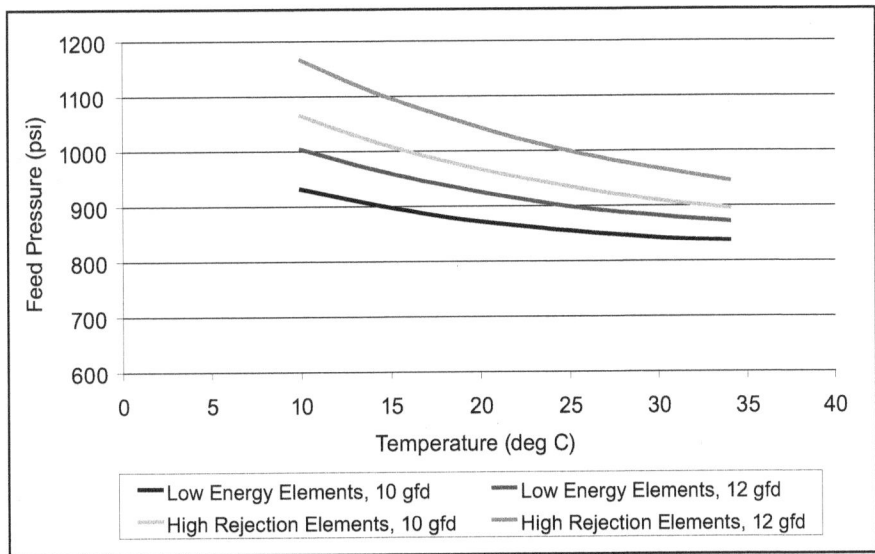

Courtesy of Ken Klinko and Greg Wetterau using Hydranautics IMSDesign (2009)

Figure 3-2 Projected SWRO feed pressure requirements as a function of influent water temperature for different flux rate and element type

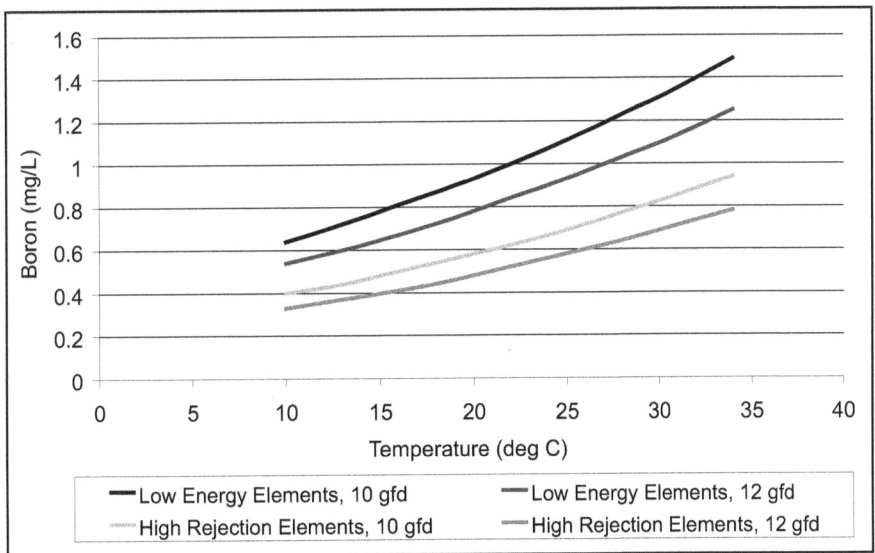

Courtesy of Ken Klinko and Greg Wetterau using Hydranautics IMSDesign (2009)

Figure 3-3 Projected impact of temperature on SWRO permeate boron

Two-Pass Nanofiltration

A novel approach to seawater desalination involving two-pass nanofiltration (NF) has been developed by the Long Beach Water Department (LBWD) in California.

In this approach, two passes of NF are used in series: the first pass removes greater than 90 percent of the TDS from the influent seawater, and the second pass removes greater than 93 percent of the remaining TDS for a total salt reduction of approximately

99.5 percent (Tseng et al. 2003). The first NF pass operates at a pressure of approximately 525 psi (3.6 MPa) and the second NF pass operates at a pressure of approximately 250 psi (1.7 MPa). These reduced operating pressures result in operational cost savings.

LBWD cites several other benefits of the two-pass NF system. Chiefly, the presence of two passes of NF implies two barriers for contaminants and therefore increased reliability of water quality. It also can increase operational flexibility; for example, the second pass can be operated at a higher pH allowing for increased rejection of boron. The concentration of boron in the Pacific Ocean typically ranges from 4 to 5 mg/L. LBWD has reported boron concentrations in the permeate to be less than 1 mg/L (i.e., the California notification level for boron) at a pH of 9.2 (Cheng 2010). Increasing the pH to 9.8 resulted in permeate boron concentrations of less than 0.5 mg/L.

The overall recovery from the process ranges from 30 percent to 45 percent for seawater desalination. LBWD has patented this process and is optimizing it in terms of membrane selection and operation.

DISINFECTION

In the U.S., disinfection requirements for drinking water supplies are established by individual states based on two guidelines established in the USEPA Surface Water Treatment Rule (SWTR): Long Term 2 Enhanced SWTR (LT2), and the Stage 2 Disinfectants and Disinfection Byproducts Rule (Stage 2 D/DBPR). These water quality issues are more fully described in Chapter 2.

Table 3-4 includes a summary of the disinfection log reduction credits given to typical treatment processes. It should be noted that disinfection credits will vary based on dose, contact time, water quality, and other operating conditions, while credits given for membrane processes will vary by state. It is assumed that seawater would typically be placed in Bin 1, based on LT2 criteria, which will not require additional treatment for *Cryptosporidium* removal or inactivation; however, it is important to consider the possibility of seawater contamination by wastewater treatment plant discharges if located close to open seawater intakes.

The California Department of Public Health (CDPH) has granted RO membranes 2-log removal credits for viruses, *Giardia*, and *Cryptosporidium* based on the continuous monitoring of 2-log TDS reduction, using online conductivity analyzers. Other surrogate measurements are being evaluated to possibly give higher removal credits; however, most states have yet to adopt formal policies on pathogen removal credits for RO membranes. Selection of the treatment process should therefore consider the requirements and specific credits given by the individual state in which the plant is operating.

Desalinated water can also impact disinfection residual management in the distribution system if chloramines are used. The primary concern is that bromide in the desalinated water will impact the initial chlorine demand and cause rapid decreases in disinfectant residuals once the water is blended with other sources in the distribution system, which may cause temporary noncompliance with regulations.

As discussed in Chapter 2, blending of desalinated water containing bromide with water from distribution systems containing treated surface water can have potential impacts on DBP formation. Desalinated seawater is low in TOC, which can serve as a precursor to DBP formation; however, when the desalinated seawater is blended with higher TOC surface water supplies, the DBP formation in the blend can be greater than in either of the waters individually. Blending studies are therefore recommended where DBP formation is a concern.

Table 3-4 Log removal credits for potential treatment processes

Pathogen	Log Removal Credits Required[1]		Log Removal and/or Inactivation Credits Allocated for Typical Treatment Processes						
	Min	Max	Slow Sand Filtration	Conventional Treatment	MF	UF	RO [4]	UV	Chemical Disinfection[5]
Virus	4	6.0[2]	1	2	0-1	0-4	2	0	2-4
Giardia	3	5.0[2]	2	2.5	4	4	2	3	0.5-3
Cryptosporidium	2	5.5[3]	2	3	4	4	2	3	0

1. States may require a minimum 0.5 log of *Giardia* or 2-log of virus disinfection beyond filtration to provide a multi-barrier treatment approach.
2. For California only, based on CDPH guidelines for impaired water sources.
3. Based on LT2 ESWTR for bin 4.
4. Removal credits for RO are based on CDPH policy, based on continuous monitoring of TDS reduction. Removal credits for RO may differ as more states adopt policies on seawater desalination.
5. Chemical disinfection refers to the use of free chlorine following the RO process. *Cryptosporidium* removal credits may be achieved utilizing chlorine dioxide or ozone.

Courtesy of Greg Wetterau

POSTTREATMENT

Because RO treatment removes most of the hardness and alkalinity of the source seawater, RO permeate must be remineralized to prevent corrosion in distribution system piping and to produce a finished water that is aesthetically acceptable to the customers. It is typically best to match existing distribution system water quality to the maximum extent possible to avoid customer complaints or problems with pipeline corrosion or release of scale in the distribution system. Such concerns are common to all RO designs; however, unlike brackish RO systems, where a percentage of raw water can be bypassed around the RO for product water stabilization, seawater RO facilities must make use of an external source of hardness and alkalinity.

In SWRO facilities, hardness is most commonly added with either calcium oxide (quicklime), calcium hydroxide (hydrated lime), or calcium carbonate (limestone or calcite). Quicklime is typically the least costly alternative, but it requires slakers that are operationally intensive and sometimes seen as undesirable by plant operators. In addition, the use of quicklime or hydrated lime at the tail end of the treatment process will likely require lime saturators to avoid problems with turbidity from inert materials and incomplete dissolution of the lime. These saturators will add to the cost and operating complexity of the lime feed system, but may be necessary to produce the high quality product water.

Calcite or limestone contactors are the simplest approach for remineralization from an operations perspective, but these require large footprints for calcium saturation to occur and can be a costly approach for large treatment facilities. Calcite contactors are common for small desalination facilities, such as the recently built 0.3 mgd (1.1 MLD) facility in Sand City, California; however, they were also employed in the 52.8 mgd (200 MLD) Barcelona facility in Spain, and are planned for the 50 mgd (189 MLD) desalination plant in Carlsbad, California. The best approach for remineralization should consider both economic factors and operations preferences for the utility.

Carbon dioxide is frequently added to provide alkalinity, and sodium hydroxide is often added to adjust the pH to match the existing pH within the distribution system. Other alternatives for product water stabilization include the use of calcium chloride with sodium hydroxide, use of orthophosphate or polyphosphate, or blending with a readily available source of hard water, such as a local surface water or groundwater.

ENERGY RECOVERY

Energy represents the single greatest cost for operating SWRO facilities due to the high feed pressure requirements. To reduce the volume of feedwater pumped, the most common approach is to increase system hydraulic recovery. Higher hydraulic recovery during the desalination process leads to more concentrated brine, which in turn has a higher osmotic pressure leading to an even higher feed pressure. Much of this pressure still remains in the concentrate waste stream, with the inherent energy sometimes wasted, being burned over an orifice plate or throttling valve downstream of the membranes. Many SWRO systems therefore include an energy recovery system to recapture the energy remaining in the concentrate stream, reducing the new energy input required for the feedwater.

Energy Recovery Devices

There are now a variety of commercially available energy recovery devices (ERD) that allow energy recovery from the brine stream to reduce electrical pumping costs. The primary types of ERD can be divided into two major categories: centrifugal devices and positive displacement devices. Centrifugal devices include the Francis Turbine, the Pelton wheel turbine, and the hydraulic turbocharger; positive displacement devices include the work exchanger and pressure exchanger.

Each type of ERD is described in the following sections.

Francis Turbine. In a Francis Turbine, water enters the turbine runner with a radial velocity component and discharges with an axial velocity component, like a reverse running pump. Francis Turbines are distinguished by having a band that surrounds the peripheral end of the blades (also known as *buckets*), providing a boundary for the water passage and structural rigidity to the runner. Francis Turbines are directly coupled to the feed pump and must be designed for specific operating conditions. The result is that changes in flow and pressure must be bypassed around the unit, lowering recovery efficiency.

Pelton impulse turbines (PIT). The Pelton wheel turbine operates by converting the velocity energy from a brine stream into kinetic energy. Nozzles aim the pressurized concentrate stream towards the Pelton wheel, which rotates a turbine, creating electrical energy to assist the electric motor in driving the high pressure feed pumps. Up to 80 percent of the concentrate energy can be recovered using this device. The initial capital cost is relatively high, because it must be incorporated into the feed pump and must employ high alloy stainless steels to be compatible with seawater concentrate. Figure 3-4 shows a photo of Pelton wheel generators installed at the Tampa Bay SWRO facility.

Hydraulic turbochargers (HTC). A hydraulic turbocharger is an integral, centrifugal feed pump with an energy recovery turbine. The HTC receives pressure from the RO concentrate and returns it to the RO system as a pressure boost to the feed stream resulting in reduced feed pressure required from the RO pressure pump. Hydraulic turbochargers tend to have peak efficiencies between 50 to 60 percent, with efficiency dropping as flow and pressure change from the primary design point. Figure 3-5 illustrates how a hydraulic turbocharger works in an RO system, and Figure 3-6 shows a picture of a hydraulic turbocharger.

Pressure exchangers. A PX™ pressure exchanger device transfers brine pressure energy directly to a portion of the incoming feedwater. A booster pump then makes up the difference required to achieve the needed feed pressure. This stream then joins the portion of the feed from the high pressure feed pumps that has not passed through the device. The PX™ pressure exchanger device has a single moving part, a shaftless ceramic rotor, which is suspended within a sleeve. PX™ devices can recover up to 98 percent of the energy remaining in the concentrate. Their high efficiency has contributed their increased use at many recently built SWRO facilities. Figure 3-7 shows a flow diagram of a PX™ device from Energy Recovery Inc. (ERI) installed on an RO skid. Figure 3-8 depicts an image of a typical PX™ Energy Recovery Device (ERD) installation.

TREATMENT APPROACHES   39

Courtesy of Robert Huehmer

Figure 3-4   Pelton wheel generators at Tampa Bay SWRO facility

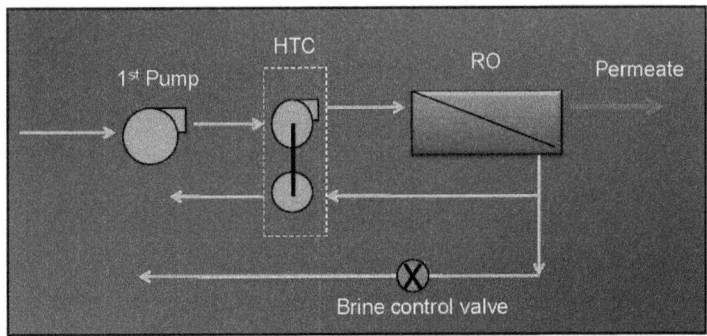

Courtesy of Val S. Frankel

Figure 3-5   Hydraulic turbocharger in an RO system

Courtesy of Energy Recovery Inc.

Figure 3-6   ERI™ TurboCharger device (low pressure turbine)

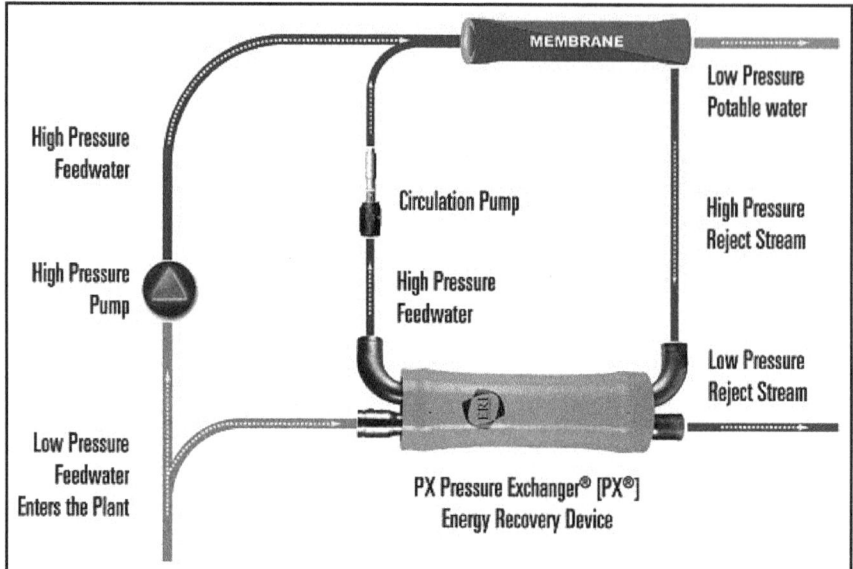

Courtesy of Energy Recovery Inc.

Figure 3-7 ERI™ PX™ energy recovery device flow diagram

Courtesy of Greg Wetterau

Figure 3-8 PX™ Pressure exchanger device installation in Sand City, Calif.

Dual work exchangers (DWEER). In a work exchanger configuration, the high pressure brine is directed to a work exchanger vessel filled with seawater to pressurize the influent seawater. A small recirculating pump boosts the pressure of the seawater exiting the work exchanger vessel to equal the main feed pump pressure and joins the flow to the membranes. Efficiencies of the work exchanger piston system can be up to 97 percent, similar to the pressure exchanger. These devices are more efficient than centrifugal designs, which rely on shaft conversion of power. Figure 3-9 shows a typical DWEER flow diagram, while Figure 3-10 depicts a DWEER installation in the Bahamas.

ERD can be installed separately on each RO train, or a multi-train concept can be applied at large SWRO plants, where both RO feed pumps and ERDs are arranged to serve multiple RO trains. The flow schematic of the 88 mgd (333 MLD) Ashkelon SWRO plant illustrates a multi-train approach shown in the Figure 3-11.

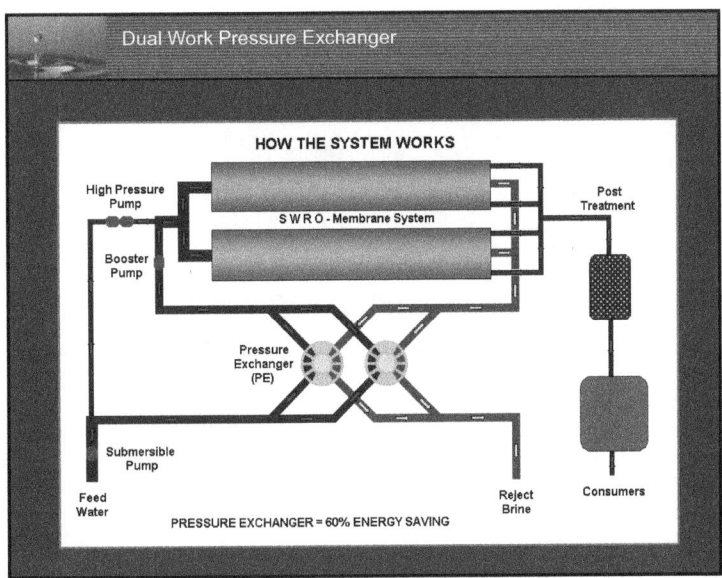

Courtesy of Val S. Frankel

Figure 3-9 Dual Work Pressure Exchanger flow diagram

Courtesy of Srinivas Veerapaneni

Figure 3-10 Installation of Flowserve DWEER energy recovery device

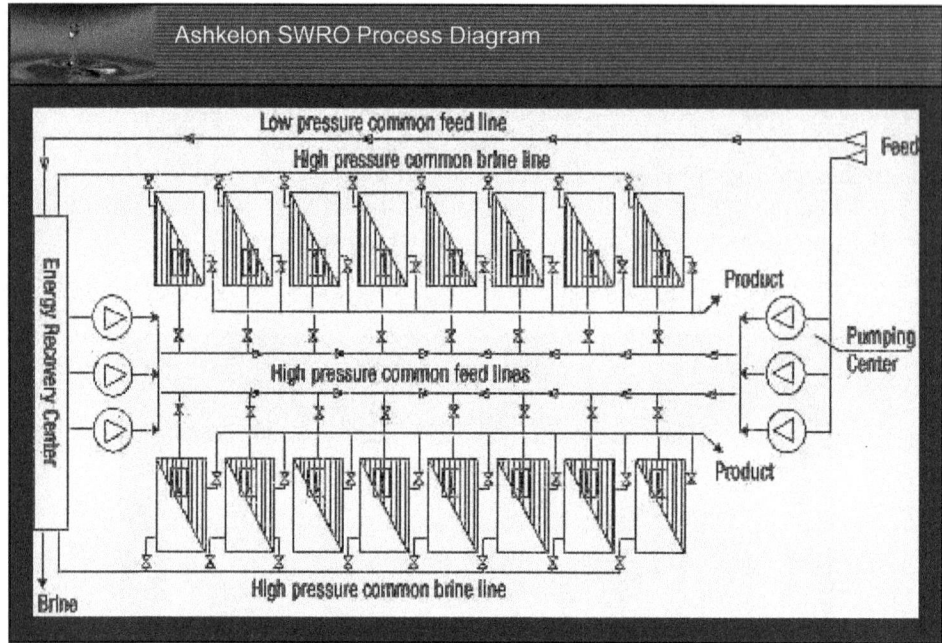

Courtesy of IDE Technologies LTD.

Figure 3-11 Three center design layout

Affordable Desalination Collaboration (ADC)

The ADC is a nonprofit organization composed of federal and state agencies, special water districts, and leading companies in the desalination industry. The goal of the ADC is to demonstrate that SWRO desalination can be a low energy-consuming and cost-effective source of freshwater in California, and to provide water agencies and private companies studying the feasibility of seawater desalination plants with opportunities for acquiring practical operations and maintenance information.

Conducted at the U.S. Navy's Seawater Desalination Test Facility in Port Hueneme, California, the ADC program effectively demonstrated the capability of SWRO technology to produce affordable potable water. The demonstration tests utilized a unique combination of proven technologies, equipment, and designs to indicate the cost-effectiveness and energy efficiency of desalination over alternate sources of water supply in the state.

The ADC achieved a world record for low energy seawater desalination by RO at 6.00 kWh/kgal (1.58 kWh/m^3), operating at a flux of 6 gfd and a recovery of 43 percent. It should be noted that these reported energy demands included only the RO process, rather than the entire treatment facility. High efficiency positive displacement pumps were also used, which have proven to be effective at small SWRO facilities (less than 1 mgd), but are less applicable for large-scale applications. While it may not always be feasible to employ all of the energy-saving measures utilized by the ADC, the lessons learned from operation of this facility provide a new energy efficiency benchmark for future full-scale designs.

Summary

Development of ERD has allowed reduced energy demand for SWRO desalination by almost three times from 27 kWh/kgal (7.1 kWh/m^3) in 1980 to as low as 10 kWh/kgal (2.6 kWh/m^3) by the early 2000s. The technologies are improving, and this tendency will continue as a result of continued research and different initiatives around the globe.

Table 3-5   Energy recovery devices (ERD): pros and cons

ERD Type	ERD Category	Max Efficiency (%)	Concentrate Mixing with Feedwater	Concentrate Repumping Required	Sidestream Booster Pump Required
Reverse Running Pumps (RRP)	Centrifugal Devices	75	No	No	No
Pelton Impulse Turbines (PIT)	Centrifugal Devices	80	No	Yes	No
Hydraulic Turbo-chargers (HTC)	Centrifugal Devices	65	No	No	No
Pressure Exchangers (PX)	Positive Displacement Devices	98	Yes	No	Yes
Dual Work Exchangers (DWEER)	Positive Displacement Devices	97	Yes	No	Yes

Courtesy of Val S. Frankel

Each commercially utilized ERD technology has certain pros and cons, some of which are summarized in Table 3-5.

CORROSION AND MATERIALS OF CONSTRUCTION

In the design of desalination facilities, potential corrosion issues require serious attention. Within an SWRO plant, a number of potential corrosion problems may occur as a result of the aggressive environment of seawater and concentrate in which the equipment must function. Nonmetallic materials are often used, wherever possible, to reduce corrosion impacts; however, the high pressures seen in SWRO facilities require that a major portion of the piping, valves, and fittings be metallic in construction. A number of factors can influence corrosion rates in metallic materials, including the chemical composition of the material, heat treatment utilized during manufacturing, fabrication techniques, surface condition, and design. A number of environmental factors can also influence corrosion, including solution composition, pH, temperature, oxygen content, flow velocity, deposited solids, and microbial growth. As a result, the design of SWRO facilities should carefully consider and evaluate the materials of construction.

Corrosion Types and Mitigation

Major types of corrosion of concern include: crevice, galvanic, pitting, microbial, under-deposit, and stress corrosion. Each type of corrosion and associated mitigation approaches are discussed in the following sections.

Crevice corrosion. Crevice corrosion occurs when local differences in oxygen concentration occur at a metal surface, typically associated with small pockets of stagnant water in voids. These voids can exist with deposits on the metal surface, such as foulants or welding slag, gaskets, lap joints, or under bolt or rivet heads.

Crevice corrosion can be mitigated by minimizing crevices or other conditions that can lead to seawater stagnation, using materials resistant to crevice corrosion (Figure 3-12), and using properly constructed butt-welds, as opposed to socket welds.

Galvanic corrosion. A number of different metals are typically used to construct a desalination system. Galvanic corrosion is an electrochemical process in which one metal corrodes preferentially when in electrical contact with a different type of metal and both metals are immersed in an electrolyte. When two dissimilar metals are physically connected, the metal with the more negative potential generally has increased corrosion

44 DESALINATION OF SEAWATER

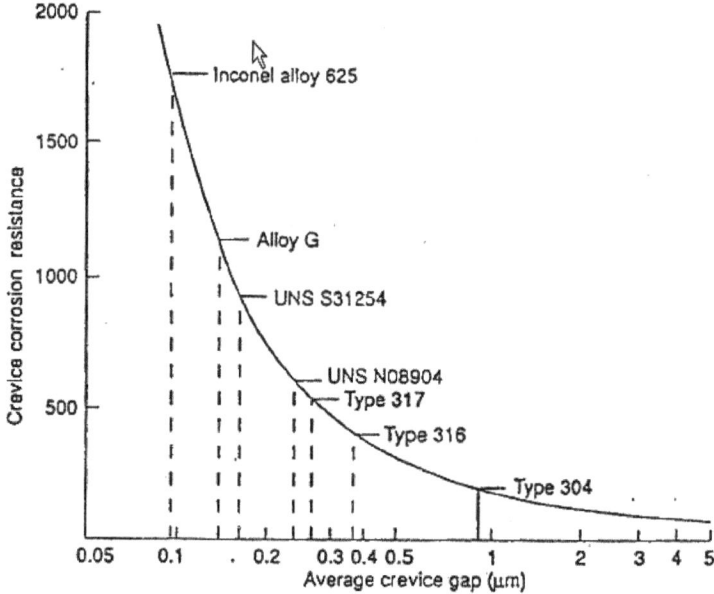

Courtesy of Canadian Institute of Mining Metallurgy and Petroleum

Figure 3-12   Resistance to crevice corrosion (Todd and Oldfield 1991)

Table 3-6   Galvanic series for alloys in flowing seawater at 4 m/s and 24°C

Material	Steady State Electrode Potential (V) v SCE
Graphite	+0.25
Platinum	+0.15
AL6XN (super austenitic stainless steel)	0.00
Zeron 100 (super duplex stainless steel)	−0.01
AISI type 316 stainless steel (passive)	−0.05
AISI type 304 stainless steel (passive)	−0.08
Hastelloy alloy C	−0.08
Titanium	−0.10
AISI type 410 stainless steel (passive)	−0.15
AISI type 316 stainless steel (active)	−0.18
Nickel	−0.20
AISI type 430 stainless steel (passive)	−0.22
Copper alloy (70Xu-30Ni)	−0.25
Copper alloy (90Cu-10Ni)	−0.28
Copper	−0.36
AISI type 304 stainless steel (active)	−0.53
Carbon steel	−0.61
Cast iron	−0.61
Zinc	−1.03

Adapted from Metal Handbook, 1987

compared to its corrosion rates when not coupled to another metal. Table 3-6 presents a galvanic series list for common metallic materials that may be used in SWRO facilities.

Galvanic corrosion can be mitigated by using materials adjacent to each other in the galvanic series, ensuring that the key component is of a more noble material, ensuring that the less noble material is present in a much larger area than the more noble material, and by using insulation to decouple dissimilar metals (e.g. nonconductive sleeves and gaskets). Simply coating or painting metal surfaces to mitigate galvanic corrosion is of limited value and should be avoided.

Pitting corrosion. Pitting corrosion is a form of extremely localized corrosion that leads to the creation of small holes in the metal. The corrosion penetrates the metal, with limited diffusion of ions, further pronouncing the localized lack of oxygen. Pits can range in size from those that are difficult to detect with the naked eye to those with a diameter and depth that can be measured in millimeters. Pitting occurs when the protective film of stainless steels breaks down in small isolated spots. Pitting corrosion is typically associated with high chloride concentrations.

Pitting corrosion can be mitigated by using chromium (Cr), molybdenum (Mo), and nitrogen (N) based alloys. One percent molybdenum has approximately the same effect as three percent chromium. Resistance to pitting corrosion is rated by a pitting resistance equivalent number (PREN)—with the higher the PREN, the higher the resistance to pitting corrosion. Equation 1 shows how the PREN is calculated where element contents are given in wt %.

$$PREN = \%Cr + 3.3(\%Mo) + 16(\%N) \qquad (Eq\ 1)$$

Typically, austenitic and duplex stainless steel piping used in desalination should have a PREN of 40 or more. Table 3-7 provides PREN values for materials commonly used in SWRO facilities.

Microbial corrosion. Microbially induced corrosion (MIC) occurs when microorganisms attach to surfaces such as pipe walls and metabolize available nutrients using dissolved oxygen or other chemical compounds to generate corrosive by-products (such as acids and sulfides). This results in corrosion of the metal, forming a pit or crevice beneath the biofilm area. Sulfate-reducing bacteria, iron and manganese bacteria, and sulfur-oxidizing bacteria have most often been associated with MIC of stainless steels.

Microbial corrosion can be mitigated by using shock chlorination and/or biocide addition; using polymeric materials; or using alloys high in Mo (> 6 percent).

Under-deposit corrosion. Deposits, usually from corrosion of metal surfaces (such as iron oxide tubercles), provide electrochemical conditions that favor additional corrosion and buildup of deposits. In the worst cases, under-deposit corrosion causes deep pitting and pipe failure. The accumulated deposits also create favorable conditions for MIC.

Under-deposit corrosion is best mitigated by preventing surface corrosion in the first place. Once corrosion has occurred on the surface, under-deposit corrosion cannot really be prevented.

Stress corrosion. Stress corrosion and associated cracking can occur in materials exposed to both high tensile stresses and specific corrosion-inducing conditions, and is temperature dependent. Stress corrosion of stainless steels in chloride containing water rarely occurs at temperatures below 70°C. This issue is typically associated with thermal desalination plants, as opposed to RO plants.

General Corrosion Mitigation Practices

The prevalence of corrosion is typically minimized by the following practices:

- Material Selection—Proper material selection for the chemical conditions present

Table 3-6   PREN values for common materials

Type	Material	UNS Number	PREN
Austenitic	304 L	S30403	18–20
	316 L	S31603	24–28
	317 L	S31703	29–31
	904 L	N08904	35–36
Super Austenitic	AL-6XN	N08367	47–48
	254 SMO	S31254	43–46
Duplex	2205	S32205	35–37
	329	S32900	34
Super Duplex	SAF 2507	S32750	42
	Zeron 100	S32760	≥ 40 (42)
Austenitic	Alloy 20 (20Cb-3)	N08020	31
	Hastelloy C-276	N10276	68
	Hastelloy C-22	N06022	65
—	90-10 Cu-Ni	C70600	NA

Courtesy of CH2M HILL

- Use of Isolation—Isolating sleeves, gaskets, etc., to decouple the metals from forming an active electrochemical cell.
- Flushing—Using appropriate flushing to minimize stagnation of high chloride water in piping. For intermittently used pipes, permeate flushes can be used.
- Flow velocity—Low flow velocities and stagnant conditions increase the corrosion potential in austenitic stainless steels (but not super austenitic stainless steels) in high chloride environments. Generally, piping velocities should not exceed 5 feet per second (fps) (1.5 m/s) upstream of pumps and not exceed 10 fps (3 m/s) in other parts of the system. High velocities may also result in scour and or impingement. Design velocities for austenitic stainless steel piping should be a minimum of 5 fps (1.5 m/s). When copper alloys are used, design per manufacturer recommendations for the specific product.
- Nonmetallic lining—As a cost-cutting factor, some vendors advocate the use of lined or painted components rather than more expensive components possessing appropriate corrosion resistance. Linings typically work well when whole but are prone to damage during construction or use. Once linings are breached, accelerated corrosion can occur, resulting in rapid failure, so the use of linings to mitigate corrosion should be considered carefully.
- Chlorination—Shock chlorination is frequently practiced to mitigate biofouling in SWRO systems but must be followed by complete dechlorination. Investigations have shown that chlorination increases the risks for localized corrosion in seawater (especially crevice corrosion). Continuously chlorinated seawater is considerably more aggressive than unchlorinated or intermittently chlorinated seawater. Increasing water temperature increases the corrosion potential in chlorinated seawater. The highest alloyed stainless steels are resistant to crevice

corrosion, but lower grade stainless steels are not. In contrast, the risk of galvanic corrosion decreases if the seawater is chlorinated.

- Ventilation—Many rotating machinery components and control system components require ventilation and cooling. Totally enclosed fan cooled (TEFC) components, for instance, rely on airflow across the windings using an integrated fan located on the end of the motor housing for cooling. Many electrical components that may accumulate heat, such as control panels, transformers, and variable frequency drives, all typically utilize fan-based cooling. Air in coastal regions typically is laden with seawater aerosols. These aerosols coat components being cooled and because they are an electrolyte, may result in corrosion, short-circuits, and electrical failures. External cooling, or clean-dry air purge system, or another appropriate method is encouraged for cooling.

- HVAC—Designers must account for seawater aerosols in the design of HVAC equipment. Dispersion of the aerosols through the HVAC system into the conditioned facilities should be discouraged. Accelerated corrosion of HVAC components exposed to seawater aerosols may exist.

- Reinforced concrete—Most concrete is somewhat porous, allowing the migration/penetration of water (and seawater) into the concrete. Galvanic corrosion of rebar can result in the release of hydrogen gas, and subsequent spalling of the concrete – substantially reducing the concrete's serviceable life. Mitigation can include some combination of the use of low permeability admixes, use of coated rebar, ensuring a minimum thickness of concrete over rebar and the use of anodic protection of the rebar systems.

- Valves—While most valves can be constructed of corrosion resistant materials, the use of dissimilar materials can result in corrosion. All of the valve trim must also be of appropriate materials. Air relief valves on seawater pipelines in desalination plants are often specified incorrectly – resulting in corrosion from trim materials.

- Instrumentation—Appropriate selection of wetted materials of instrumentation is required. If not available, an isolating style diaphragm may be required to protect the instruments. Those most prone to failure are pressure instruments and orifice plate or vortex style flow meters.

- Lined materials—There has been an increasing trend to use plastic components as frequently as possible in desalination facilities. As a result, the industry is seeing an increased use of reinforced plastic for piping, pressure filter vessels, pumps, cartridge filter vessels, and other components. It is anticipated that this trend will accelerate.

REFERENCES

American Water Works Association. 2007. M46 *Reverse Osmosis and Nanofiltration.* Denver, Colo.: American Water Works Association.

Chave K.E. and E. Suess. 1970. Calcium Carbonate Saturation in Seawater: Effects of Dissolved Organic Matter. *Limnology and Oceanography* 15(4): 633-637 American Society of Limnology and Oceanography.

Cheng, R., C. Andrews-Tate, T. Tseng, and K. Wattier. 2010. Distribution of Desalinated Seawaters: Are Corrosion Indicators Sufficient. *IDA Journal*, First Quarter 2010.

IDA Desalination Yearbook 2009-2010, www.desalyearbook.com/about.html

Metal Hand Book Vol. 13, *Corrosion*, ASM 1987.

Todd, B. and J.W. Oldfield. 1991. "Reverse Osmosis—Which Stainless Steel to Use." *Corrosion Management* (acorn) No. 1-2.

Tseng, T., R. Cheng, D. Vuong, K. Wattier. 2003. Optimization of Dual-Staged Nanofiltration Membranes for Seawater Desalination. Proc. of the AWWA 2003 CA–NV Annual Fall Conferences.

Voutchkov, N. 2008. "Pretreatment Technologies for Membrane Seawater Desalination" Sydney, Aus.: Australian Water Association.

Wilf, M. and K. Klinko. 2001. Optimization of seawater RO system design. *Desalination* 138: 299–306.

Chapter 4

Environmental Impacts and Mitigation Measures

Nikolay Voutchkov
Henry Hunt

INTRODUCTION

The purpose of this chapter is to provide an overview of key environmental impacts of seawater desalination plant construction and operation, and to discuss alternatives for environmental impact minimization and mitigation.

The environmental impacts of seawater desalination plant operations have many similarities to those of conventional water treatment plants. Similar to conventional water treatment facilities, desalination plants have source water intake and waste stream discharge that may impact the aquatic environment in which they are located. In addition, desalination facilities and conventional water treatment plants may use many of the same chemicals for source water conditioning, and therefore, have similar waste streams, apart from salinity, associated with the disposal of the spent conditioning chemicals and the source water solids. Seawater desalination plants, however, use large pumps and motors that have potential to be larger sources of noise pollution than similarly sized conventional plants. These pumps also consume relatively large amounts of electricity and therefore, may have direct and indirect impacts on air quality and greenhouse gas emissions.

Despite many of the similarities of their environmental impacts, desalination plants have several distinctive differences as compared to conventional water treatment plants: (1) they use approximately 1.5 to 2.5 times more source water to produce the same amount of fresh water; (2) they generate a discharge with elevated salinity, which typically has

1.5 to 2 times higher TDS concentration than that of the source seawater; and (3) they use five to ten times more electricity for treatment of the same volume of freshwater.

The environmental impact of desalination plant operations should be assessed in the context of the environmental impacts of water supply alternatives that may be used instead of desalination. Desalination projects are typically driven by the limited availability of alternative lower-cost water supply resources, such as groundwater or fresh surface water (rivers, lakes, etc.). However, environmental impacts may also result from continuation of those water supply practices. For example, over-pumping of freshwater coastal aquifers for years in a number of areas has resulted in a significant increase in the salinity of the groundwater and has damaged these aquifers. In some arid areas, transfers of fresh water from a traditional water supply source, such as a river, river delta, or a lake, have impacted the eco-balance in this freshwater source to an extent that the long-term continuation of this water supply practice may result in significant and irreversible damage of the ecosystem of the traditional freshwater supply source. In such cases, the environmental impacts of the construction and operation of a new seawater desalination project should be weighed against the environmentally damaging consequences from the continuation/expansion of the existing fresh-water supply practices. In addition, the impacts of a seawater desalination facility should be considered against the impacts of water reuse alternatives, both potable and nonpotable.

Waste streams generated from desalination plants, with the exception of the high-salinity reject water, are similar to the waste streams generated by conventional water treatment plants and water reuse facilities. Water reclamation plants also generate waste streams that contain some of the same chemicals used for desalination and may also have elevated content of man-made waste substances, which may have potential impacts on the marine environment.

SOURCE WATER INTAKES

The main purpose of intakes is to collect source seawater of adequate quantity and quality needed to produce desalinated water. Because intake water quality has a significant impact on desalination plant operations, desalination intake design should target collection of water with minimal inorganic, to the extent possible, and organic content, including marine life. As indicated in Chapter 3 of this manual, seawater desalination plants use two types of source seawater collection facilities: subterranean or subsurface intakes (wells and infiltration galleries) and open ocean intakes. It should be noted that a subsurface infiltration gallery will typically operate like a well intake with respect to entrainment and impingement issues. However, it may face similar impacts during construction as an open intake, because construction is typically done offshore and because of the large area needed for this intake.

Impingement and entrainment of marine organisms by the desalination plant intake are considered the two main potential environmental impacts of these facilities, and are particularly associated with open ocean intakes. Impingement occurs when aquatic organisms are trapped against intake screens by the velocity and the force of the flowing source water. Entrainment occurs when marine organisms pass through the intake screens and enter into the process equipment and treatment facilities where some of them are destroyed.

The impacts of impingement and entrainment vary considerably with the volume and velocity of feed seawater and the use of mitigation measures developed to minimize their impact. Impingement and entrainment of aquatic organisms are not environmental impacts unique to open intakes of seawater desalination plants only. Conventional freshwater open intakes from surface water sources (i.e., rivers, lakes, estuaries) may also cause measurable impingement and entrainment. Often, freshwater sources contain a large content and variety of aquatic species, similar to open ocean waters. However, the impingement and entrainment impacts of these intakes have been either accepted or addressed at numerous freshwater supplies throughout the United States. Disproportionately elevated attention of impingement and entrainment issues associated with seawater intakes may stem, in part, from federal regulations that address this topic for power generation plants and from the environmental scrutiny associated with their public review process.

Similar to environmental impacts from other aspects of desalination plant operation, the magnitude of impacts due to entrainment and impingement varies significantly from one location to another. Therefore, when assessing the impacts caused by the intake of a desalination facility, it is essential to consider the applied technology and operational practices, the actual volumes and velocity of water being drawn into the desalination plants, and the species composition and abundance of the seawater surrounding the intake.

Subterranean or Subsurface Intakes—environmental impacts and mitigation measures.

Subsurface intakes could have a number of environmental impacts, such as loss of coastal habitat during construction, visual and aesthetic impacts, and impacts on nearby coastal wetlands depending on their method of construction and their design for well completion. The magnitude of these impacts and potential mitigation measures are discussed in the following sections for the installation of subsurface intakes constructed as wells, commonly referred to as beach wells. Such impacts and widely used mitigation measures are also discussed.

Impingement and entrainment. Because subsurface intakes naturally filter the collected seawater at low velocities through the granular formations of the coastal aquifer in which they operate, their use minimizes entrainment of marine organisms into the seawater desalination plant. It should be noted however, that to date no scientific or engineering studies have been performed to assess and document the entrainment impact of subsurface intakes because usually regulatory agencies assume that such impact is insignificant. The source seawater collected by this type of intake typically does not require mechanical screening, and therefore, subsurface intakes do not cause impingement impacts on the marine organisms in the area of the intake.

Visual and aesthetic impacts and mitigation measures. The visual and aesthetic impacts of beach well intakes are dependent on the location of the wellhead and the style of well completion used. If the beach intakes (wells) can be constructed below grade, at grade, or near grade to minimize impacts, submersible well pumps can be installed below grade and the structures made watertight. The electrical controls and auxiliary equipment can be installed within the watertight structure or located at a remote location near the intake, off the beach, for protection. In these cases, there may be little or no visual or aesthetic impacts for this kind of intake completion.

If the beach intake must be constructed above grade (see Figure 4-1), the magnitude of this impact will vary according to the physical placement of the wells and the height above grade that is required. With radial collector wells, it is possible to locate the well structure back from the beach and extend the well screens out underneath the beach to reduce visual impacts.

Considering that the desalination plant source water must be protected from acts of vandalism and terrorism, the individual beach wells may have to be fenced off or otherwise protected from unauthorized access (see Figure 4-2).

The larger beach well (e.g., concrete) must have secured access and/or be fenced off, which damage's the beaches visual and aesthetic appeal, while subgrade or near-grade completion could utilize secured access hatches and would have limited impacts. Because beaches are visually sensitive areas, the installation of above-grade beach wells may affect the recreational and tourism use and value of the seashore, and may change the beach appearance and character if structures cannot be located at strategic locations within the area.

52   DESALINATION OF SEAWATER

Courtesy of Water Globe Consulting

Figure 4-1   3.8 MGD intake beach well of a large seawater desalination plant

Courtesy of Water Globe Consulting

Figure 4-2   Beach well intake system (abovegrade completion)

For comparison, open coastal intakes that can have the pumping facilities located back from the shoreline are typically lower-profile structures that may blend better with the coastal environment and its surroundings. However, if a large pumping structure is needed to house numerous pumps and/or screening systems, even a well-set-back structure, whether open intake or beach well, may have visual impacts on the environment.

Installing the intake wells and pumping gallery in a set-back location, often located behind the beach, is usually preferable, especially if less environmentally sensitive area of adequate size is available near the desalination plant site and the shore (see Figures 4-3 and 4-4). These two general locations for the wells can utilize different well designs to accommodate local geographic settings and other social-environmental issues. These designs include wells that are:

  a. Completed below grade, which can include the wellhead being completely buried to eliminate visual impacts, either on or behind the beach.

  b. Completed at or near grade with only minimal surface features to provide low visual impacts for locations in public use and residential areas (Figure 4-3).

  c. Completed above grade, especially where the top of the well structure needs to be above known or anticipated flood elevations, and to allow access during high water events, on or behind the beach (Figure 4-4).

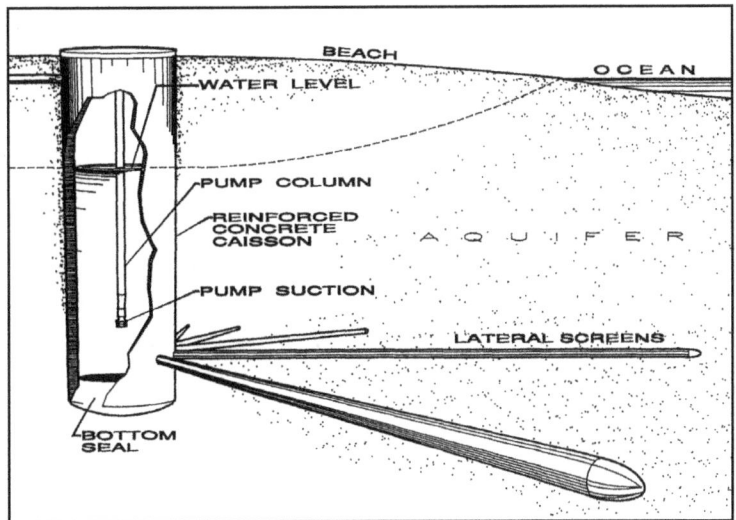

Courtesy of Ranney Collector Wells – Layne Christensen

Figure 4-3   Beach well intake system (at grade completion)

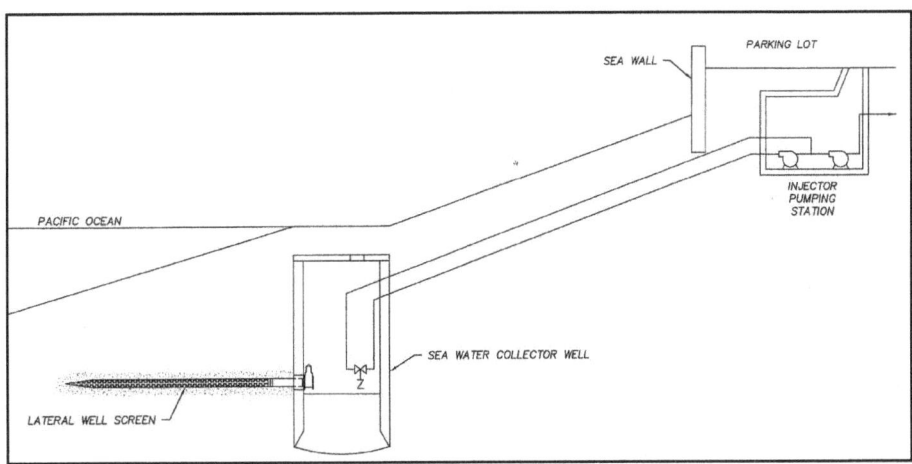

Adapted from Rando & Brady 1966

Figure 4-4   Beach well intake system (dual completion)

d. Completed with a dual-design, whereby the well portion of the system can be located closer to the source water (e.g., out on the beach), and the pumping portion of the system can be located further back from the source water to minimize impacts, typically connected with underground piping (Figure 4-4).

It should be noted that the use of the more environmentally palatable intake well configurations shown on Figures 4-3 and 4-4 will result in increase of the overall costs for intake construction.

Loss of coastal habitat during construction—impacts and mitigation measures. Smaller seawater desalination plants typically require a limited number of intake wells, and their impact on the coastal habitat during construction is generally minimal. These lower capacity wells can often be constructed as low-profile structures to minimize visual impacts. Because of the higher number of wells needed to supply adequate amounts of water for a large seawater desalination plant, construction of these facilities may result in impacts over a larger area of coastal habitat, and because these structures are often constructed as above-grade structures, they have more visual and aesthetic impacts.

Due to the increased size of the impacted seashore area for larger intake well systems, use of beach wells may result in another site-related implication, i.e., encountering artifacts of historical and archeological significance. At many locations worldwide, the probability of discovering remains of ancient habitats along the seashore is much higher than further inland as coastal or "near-water" settings were often the site of previous communities. This probability would increase with increasing the footprint of the disturbed seashore area.

Coastal wetland habitat—impacts and mitigation measures. Any intake wells that are operated in coastal areas will likely have impacts on local groundwater resources and other features such as perched water, wetlands, or saltwater-freshwater interfaces as the hydraulic influence from pumping in the area will affect water levels and alter groundwater flow gradients. Special attention should be given to seawater intake well sites in the vicinity of existing coastal wetlands and other groundwater users to evaluate these hydraulic impacts. The operation of large intake wells located adjacent to coastal wetlands may result in a drawdown of the groundwater table that could affect (dry up or destroy) the wetland habitat or impact local groundwater quality (e.g., salinity). A potential mitigation measure in this case is installation of a higher number of smaller capacity wells, where the radius of influence does not reach the wetlands, or the use of a Ranney well-type configuration. Similar concern and solution can be applied to conditions where the radius of influence of the intake wells extends to the area of landfill or other contaminated site (i.e., leaking fuel storage tanks of gas station) located near the coast. In this case, the subsurface intake could immobilize hazardous compounds contained in the hazardous waste site and contaminate the water source. Voutchkov (2004) discusses additional key factors that influence the feasibility of using subsurface intakes.

Subsurface intake construction—impacts and mitigation measures. The permanent construction-related impacts are mainly associated with the excavation and disposal of sand and other materials from the shoreline in order to drill the intake wells. From this perspective, the infiltration galleries would have the highest impact on the ocean bottom habitat, because their construction involves removing 6 to 8 feet of the ocean bottom habitat and replacing it with artificial sand and gravel. The area of the ocean bottom habitat that will be removed and destroyed is significant, especially for construction of intakes for large seawater desalination plants.

Open Ocean Intakes—environmental impacts and mitigation measures.

Similar to subsurface intakes, open ocean intakes would have environmental impacts associated with their construction and operation.

Impingement and entrainment impacts and mitigation measures. As indicated previously, impingement and entrainment are considered the two most significant environmental impacts of open ocean intakes. Impingement rates from a desalination plant open ocean intake depend on the intake design, location, and the velocity of the feedwater. Impingement mortality of marine species is typically caused by starvation, exhaustion or injury caused by the suction force of the water, or from the physical force of water jets used to clear the screens of debris.

While specific intake design may be able to reduce or eliminate impingement, all desalination open water intake systems will cause a certain degree of entrainment. Entrainment impact is associated with marine species mortality caused by the equipment, chemicals, or treatment facilities used for water treatment.

Entrainment impact is typically proportional to the volume of source water collected by the intake and varies widely based on the amount of seawater required by the facility; intake velocity; location; depth; existing biological conditions of the affected area of the intake structure; and the intake technology/equipment used. To predict and assess impacts from a desalination plant intake, site-specific studies are necessary to identify habitats and species in the area that might be vulnerable to impingement or entrainment.

The methods for mitigation of impingement and entrainment of marine organisms can be divided in three categories:

- Alternative Open Intake Technologies;
- Operational Impingement Reduction Measures;
- Impact Mitigation Measures.

Alternative desalination plant open intake technologies. Table 4-1 presents a number of technologies that are classified based on biological effectiveness (i.e., ability to achieve significant reductions in both impingement and entrainment).

The feasibility of these technologies for the site-specific condition of a given desalination project should be evaluated based on the following criteria:

- Ability to achieve a significant reduction in impingement and entrainment (IM&E) for all species, taking into account variations in abundance of all life stages;
- Feasibility of implementation at the desalination plant;
- Cost of implementation (including installed costs and annual O&M costs);
- Impacts during desalination plant operations.

Operational measures. Operational mitigation measures are used to reduce the amount of flow and velocity of entrance of the source water into the desalination plant intake to minimize entrainment and impingement of marine organisms.

Operational measures may consider reduction of plant intake flow during certain periods of the day (typically at night) and/or of the year (typically during the summer and spring months) when the concentration of marine species in the source water is at its highest levels.

Plant intake flow may be reduced by either reduction of desalination plant overall fresh water production yield and/or by operating the desalination plant at higher recovery.

Entrainment of marine organisms is mainly proportional to intake flow. Therefore, installation of variable frequency drives (VFDs) on the intake pump motors would also reduce the flow that enters the desalination plant by collecting only as much flow as needed at any given time to meet the desalination plant freshwater production target.

Impact mitigation measures. In addition to the implementation of technological and operational measures to minimize impingement and entrainment impacts, the effect of these impacts on the surrounding aquatic environment can be mitigated by implementing projects aimed to preserve, restore, or enhance this environment by creating additional habitat for species in kind to the impacted marine organisms.

56 DESALINATION OF SEAWATER

Table 4-1   Potential impingement/entrainment reduction technologies

Technology	Impact Reduction Potential	
	Impingement	Entrainment
Modified traveling screens with fish return	Yes	No
Replacement of existing traveling screens with fine mesh screens	Yes	Yes
New fine mesh screening structure	Yes	Yes
Cylindrical wedge-wire screens – fine slot width	Yes	Yes
Fish barrier net	Yes	No
Aquatic filter barrier (e.g., Gunderboom)	Yes	Yes
Fine mesh dual flow screens	Yes	Yes
Modular inclined screens	Yes	No
Angled screen system – fine mesh	Yes	Yes
Behavior barriers (e.g., light, sound, bubble curtain)	Maybe	No
Variable speed drives	Yes	Yes

Courtesy of Water Globe Consulting

Mitigation projects that should be considered will target the generation or restoration of a coastal habitat comparable to that impacted by the intake. Key eligibility criteria for such mitigation projects may include

- Consistency with the applicable requirements of federal, state, and local agencies that have jurisdiction over coastal habitat restoration actions.
- Restoration of marine habitat similar to the marine habitat impacted by the intake operations.
- Projects located in close vicinity and preferably in the watershed near the intake.
- Projects that hold the promise for long-term environmental enhancement benefits.
- Projects that have opportunities for leveraging of funds/availability of matching funds.

Examples of types of mitigation projects include:

- Wetland restoration
- Coastal lagoon restoration
- Restoration of historic sediment elevations to promote reestablishment of eelgrass beds
- Marine fish hatchery enhancement
- Contribution to a marine fish hatchery stocking program
- Artificial reef development
- Kelp bed enhancement

Selection of the most suitable mitigation measures would need to be completed based on a life cycle cost-benefit analysis.

Open intakes—construction impacts and mitigation measures. Open intakes can generally be divided into two types—onshore and offshore. Construction of onshore open intakes involves minimum disturbance of marine life in the vicinity of the intake but they are often highly visible structures with potential impacts on beach aesthetics. Offshore intakes are typically constructed by installing intake pipeline directly on the surface of the ocean bottom and securing the pipeline with weighted blocks; by installing the intake pipeline in an excavation trench; or by directional drilling of the intake pipeline/tunnel under the ocean floor. Intake pipeline installation in a trench excavated from the ocean bottom usually is the most environmentally intrusive. Therefore, if the intake area contains environmentally sensitive habitats, the preferred method of intake pipeline installation is directional drilling 5 to 15 feet (1.5 to 4.5 meters) under the ocean floor. While the onshore open intake is lowest in cost, it is the most visible structure-wise, and often for this reason it is avoided. Intake structure drilled under the ocean floor is the most costly and complex type of such facility, but has the advantage of minimal disturbance of the ocean flora and fauna during construction.

CONCENTRATE DISCHARGE

One of the key limiting factors for the construction of new desalination plants is the availability of suitable conditions and locations for disposal of concentrate or concentrate stream.

Introduction

Concentrate is generated as a by-product of the separation of the minerals from the source water used for desalination. This liquid stream contains most of the minerals and contaminants of the source water and pretreatment additives in concentrated form. The concentration of minerals and contaminants in the concentrate from seawater desalination plants is usually 1.5 to 2.5 times of that in the source water depending on the recovery of the desalination plant. If chemical pretreatment is used, such as coagulants, antiscalants, polymers, or disinfectants, some or all of these chemicals may be disposed of along with the plant discharge concentrate.

The quantity of the concentrate is largely a function of the plant recovery, which in turn is highly dependent on the TDS concentration of the source water. Seawater desalination plant recovery is typically limited to 40 to 65 percent. The TDS level of concentrate from seawater desalination plants usually is in a range of 65,000 to 85,000 mg/L, while that from brackish plants may vary between 1,500 mg/L and 25,000 mg/L. The amount of particles, total suspended solids (TSS), and biochemical oxidation demand (BOD) in the concentrate is usually below 5 mg/L because these constituents are removed by the plant's pretreatment system. However, if plant pretreatment waste streams are discharged along with the concentrate, the blend may contain elevated turbidity, TSS, and occasionally BOD. Acids and scale inhibitors added to the desalination plant source water will be rejected in the concentrate and will impact its overall mineral content and quality. Often scale inhibitors contain phosphates or organic polymers.

Because membranes are more permeable to some chemicals than others, variable concentration factors may apply for specific chemicals. Exactly how the concentrate concentration factor impacts the disposal of concentrates depends heavily on the means of disposal. In some cases, volume minimization (high concentrate concentration factor) will be preferred, whereas in cases where the concentrate is to be discharged to waterways, low concentration may be more important than low volume.

For example, the Perth Seawater Desalination Plant in Australia is a two-stage RO plant operating with a first pass recovery of 45 percent and a second pass recovery of 90 percent. This corresponds to an overall concentrate concentration factor of approximately 1.7 times. Based on a source water TDS of 33,000-37,000 mg/L, the plant produces an overall RO concentrate TDS of approximately 65,000 mg/L.

With most seawater desalination plants producing concentrate 1.5 to 2 times more concentrated than ambient seawater, the concentrate may have a negative impact on the aquatic environment in the area of the discharge. This impact is very site-specific and depends mostly on the salinity tolerance of the specific marine organisms inhabiting the water column and benthic environment influenced by the discharge. The existing USEPA whole effluent toxicity (WET) tests are indicative of the level of salinity that causes mortality of preselected test organisms, which may or may not inhabit the discharge area. WET testing is an important element of the comprehensive evaluation of the effect of the concentrate discharge on the aquatic life. Completion of both acute and chronic toxicity testing is recommended for the salinity levels that may occur under worst-case combination of conditions in the discharge (Voutchkov 2006).

Mechanisms of Concentrate Impact on the Environment

Concentrate from seawater desalination plants using open ocean intakes generally has the same color, odor, oxygen content and transparency as the source seawater from which it was produced, and an increase or decrease in salinity will not change its physical characteristics or aesthetic impact on the environment.

There is no relationship between the level of salinity and biological or chemical oxygen demand of the desalination plant concentrate. More than 80 percent of the minerals that encompass concentrate salinity are sodium and chloride, and they are not a prime food source or macro- or micronutrients for aquatic organisms.

Salinity contained in concentrate discharges from seawater desalination plants is not of anthropogenic origin as are the pollutants contained in discharges from industrial or municipal wastewater treatment plants or water reclamation plants. The minerals contained in the seawater desalination plant concentrate discharge originated from the same source to which they usually are returned. As a result, the environmental effect of seawater desalination on the ocean is somewhat equivalent to the effect of naturally occurring evaporation.

Naturally occurring evaporation tends to concentrate salinity in shallow nearshore ocean embayments during the high-temperature dry periods of the year, and they are diluted during the rainy periods of the year keeping a net zero sum salinity effect. Similarly, seawater desalination plants temporarily remove a small portion of ocean water producing fresh drinking water, which in turn may be returned to the ocean via the ocean discharges of the wastewater treatment plants located in the vicinity of the desalination plant. Even at locations where extensive water reuse projects are utilized, a portion of the water will almost universally be returned to the ocean with salinities lower than the background salinity in the ocean.

Salinity Tolerance of Marine Organisms

Environmentally safe disposal of the concentrate produced at seawater desalination plants is one of the key factors determining the viability, size, and costs of a given project. The maximum total dissolved solids (TDS) concentration that can be tolerated by the marine organisms living in the desalination plant outfall area is defined as a *salinity tolerance threshold* and depends on the type of the aquatic organisms inhabiting the area of the discharge and the period of time these organisms are exposed to the elevated salinity (Voutchkov 2006). These conditions are very site-specific for the area of each desalination outfall, and therefore, it is very difficult to determine the salinity tolerance threshold.

Marine organisms have varying sensitivity to elevated salinity. Some marine organisms are *osmotic conformers*, meaning that they have no mechanism to control osmosis, therefore their cells conform to the same salinity as their environment. A large increase in salinity in the surrounding marine environment due to concentrate discharge can cause water to leave the cells of these organisms, which eventually leads to cell dehydration that can result in cell death.

Osmotic regulators are marine organisms that can naturally control the salt content, and hence control the osmotic potential within their cells, despite variations in external salinity. Most marine fish, reptiles, birds, and mammals are osmotic regulators and employ a variety of mechanisms to control osmosis. Salinity tolerances of marine organisms vary, but few shellfish (scallops, clams, oysters, mussels, or crabs) or reef-building corals are able to tolerate salinities greater than 40,000-45,000 mg/L (D.A. Lord & Associates 2005).

Concentrate disposal may also have impacts other than direct changes in salinity. In some circumstances, concentrate plume density may lead to increased stratification reducing vertical mixing (Van Senden and Miller 2005). This may reduce dissolved oxygen levels in the water column or bottom of the ocean in the area of the discharge, which may have ecological implications. For example, stratification was raised as a particular concern during the planning and assessment for the Perth Seawater Desalination Plant discharge into Cockburn Sound in Australia. However, detailed modeling and site investigation concluded that the anticipated concentrate discharge was unlikely to contribute to the exacerbation of low-oxygen conditions in this case (D.A. Lord & Associates 2005). An ongoing dissolved oxygen monitoring program has been instituted since commissioning of the plant in 2006, and the data to date indicate that the desalination plant discharge has no measurable impact on the dissolved oxygen concentration in the area of the discharge or in Cockburn Sound as a whole. Although a drop in dissolved oxygen was observed, it was found that this change was not linked to the concentrate outfall.

A comprehensive study on the effect of the disposal of seawater desalination plant discharges on near-shore communities in the Caribbean was completed in 1998 by the Southwest Florida Water Management District and the University of South Florida (Hammond et al. 1998). This study undertook a detailed analysis of the environmental impacts of the discharges from seven existing seawater desalination plants in the Caribbean with plant capacities between 0.05 mgd (0.2 MLD) and 1.6 mgd (6 MLD) and discharge salinities between 45,000 mg/L and 56,000 mg/L. All of the plants use SWRO technology for salt separation and had been in operation for at least four years before the study was completed. The study found no statistically significant impact of the desalination plant discharges on the benthic marine life, seagrass, microalgae, and micro- and macroinvertebrates inhabiting the area of the discharge.

Numerous concentrate alternatives are utilized for both brackish and seawater desalination facilities. According to a study by the Bureau of Reclamation (Mickley 2006), the concentrate disposal methods most widely used in the U.S. are those shown in Table 4-2. These results are based on a survey completed in year 2000 of 203 desalination plants. The survey included only plants with a capacity larger than 0.05 mgd (0.2 MLD). Approximately 95 percent of the surveyed plants were nanofiltration or brackish water, with only a small portion representing seawater facilities.

Direct Discharge Through New Ocean Outfall—environmental impacts & mitigation measures.

Discharge of seawater desalination plant concentrate through a new ocean outfall is widely used for projects of all sizes. More than 90 percent of the large seawater desalination plants currently in operation dispose of concentrate through a new ocean outfall specifically designed and built for that purpose. Examples of large RO seawater desalination plants with ocean outfalls for concentrate discharge are the 86 mgd plant in Ashkelon, Israel (Figure 4-5); the 36 mgd (136 MLD) Tuas Seawater Desalination Plant in Singapore; the 14 mgd (53 MLD) Larnaka Desalination Facility in Cyprus; and most of the large plants in Spain, Australia, and the Middle East.

Table 4-2 Concentrate disposal methods for existing desalination in the U.S. (including brackish RO, NF, and SWRO)

Concentrate Disposal Method	Frequency of Use (% of Surveyed Plants)
Surface Water Discharge	45
Sanitary Sewer Discharge	27
Deep Well Injection	16
Spray Irrigation	8
Evaporation Ponds	0
Others	4

Courtesy of Water Globe Consulting

Courtesy of Water Globe Consulting

Figure 4-5 Tidal zone (onshore) discharge of the Ashkelon SWRO Plant, Israel

The main purpose of every ocean outfall is to dispose of the plant concentrate in an environmentally safe manner, minimizing the size of the zone of discharge in which the salinity is elevated outside of the typical range of tolerance of the marine organisms inhabiting the discharge area. The two key options available to accelerate concentrate mixing from an ocean outfall discharge is to either rely on the naturally occurring mixing capacity of the tidal (surf) zone or to discharge the concentrate beyond the tidal zone and to install diffusers at the end of the discharge outfall to improve mixing. Although the tidal zone carries a significant amount of turbulent energy and usually provides much better mixing than the end-of-pipe type of diffuser outfall system, this zone has a limited capacity to transport and dissipate the saline discharge load into the open ocean. If the mass of the saline discharge exceeds the threshold of the tidal zone's salinity load transport capacity, the excess salinity will begin to accumulate in the tidal zone and could ultimately result in a long-term salinity increase in this zone beyond the level of tolerance of the aquatic life. Therefore, the tidal zone is usually a suitable location for discharge only when it has adequate capacity to receive, mix, and transport the salinity discharge from a desalination plant into the open ocean.

This salinity threshold mixing/transport capacity of the tidal zone can be determined using hydrodynamic modeling. If the desalination plant TDS discharge load is lower than the tidal zone threshold mixing/transport capacity, concentrate disposal to this zone is preferable and is much more cost effective than the use of a long open outfall equipped with a diffuser system. An example of onshore discharge in the tidal zone is that of the Ashkelon seawater desalination plant (Figure 4-5).

For small plants (i.e., plants with production capacity of 0.1 mgd [0.4 MLD] or less), the ocean outfall is usually constructed as an open-ended pipe that extends to 300 feet (91 meters) into the tidal zone of the ocean. This type of discharge usually relies on the mixing turbulence of the tidal zone to dissipate the concentrate and to quickly bring the discharge salinity to ambient conditions.

The majority of ocean outfalls for large seawater desalination plants extend beyond the tidal zone. Large off-shore ocean outfalls are usually equipped with diffusers in order to provide the mixing necessary to prevent the heavy saline discharge plume from accumulating at the ocean bottom in the immediate vicinity of the discharge. The length, size, and configuration of the outfall and diffuser structures for a large desalination plant are typically determined based on hydrodynamic or physical modeling for the site specific conditions of the discharge location.

A recent example of an ocean outfall in the open ocean (outside of the tidal zone) is the outfall of the 38 MGD Perth seawater desalination plant. The Perth desalination plant outlet is 48-in. (1.2 m) in diameter and has a 530-ft (160 m) long, 40-port diffuser where the ports are spaced at 16.5 ft (5 m) intervals with an 8-in. (0.22 m) nominal port diameter, located 1,550 ft (470 m) offshore, at a depth of 33 ft (10 m), adjacent to the plant in Cockburn Sound (Crisp 2007) (Figure 4-6).

The diffuser is a bifurcated double-T arrangement and incorporates a discharge angle of 60°. This design was adopted with the expectation that the plume would rise to a height of 28 ft (8.5 m) before beginning to sink due to its elevated density. It was designed to achieve a plume thickness at the edge of the mixing zone of 8.25 ft (2.5 m) and, in the absence of ambient crossflow, 132 ft (40 m) laterally from the diffuser to the edge of the mixing zone (see Figure 4-7).

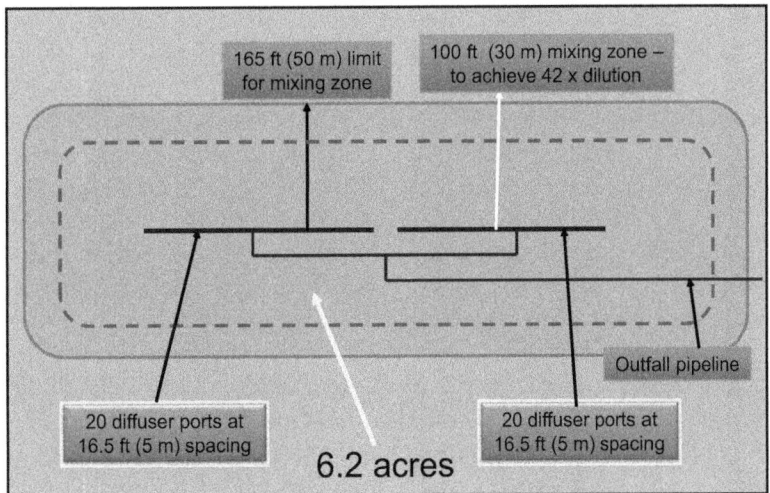

Courtesy of Water Corporation of Western Australia

Figure 4-6 Perth SWRO Plant discharge configuration

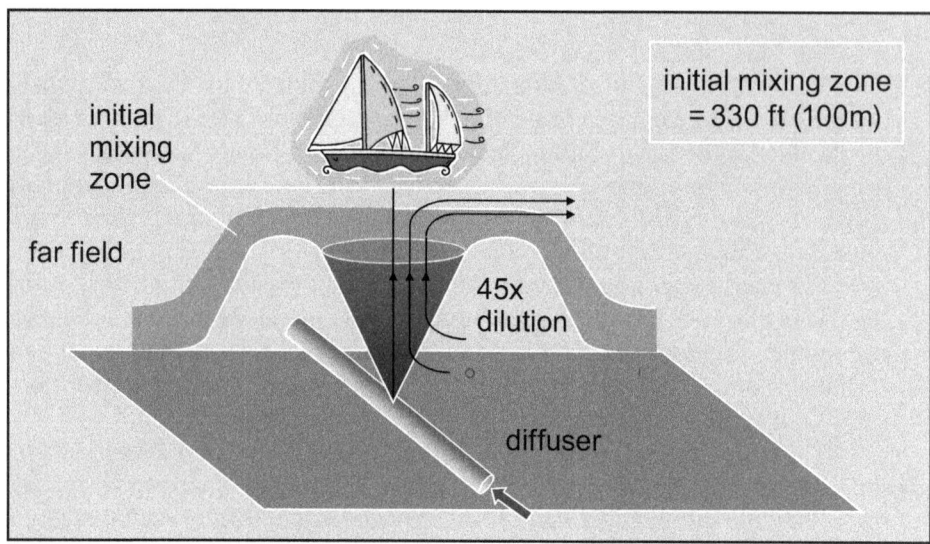

Courtesy of Water Corporation of Western Australia

Figure 4-7 Perth desalination plant mixing zone

The discharge permit for the Perth desalination plant requires that certain dissolved oxygen levels are met in order for the plant to operate. Furthermore, a minimum of 45-to-one dilution must be achieved at the edge of the mixing zone, defined in terms of a 165 ft (50 m) distance from the diffuser (GWA 2007).

Extensive real-time monitoring was undertaken in Cockburn Sound for the first year of operations (2006) to ensure the model predictions were correct and that the marine habitat and fauna were protected (Rhodes 2006). The levels of dissolved oxygen were monitored using sensors on the bed of the sound. Visual confirmation of the plume dispersion was achieved by the use of 13.7 gal (52 L) of rhodamine dye added to the plant discharge. The expulsion of the rhodamine dye from one of the plant diffusers is shown in Figure 4-8.

The dye was reported to have billowed to within about 10 ft (3 m) of the water surface before falling to the seabed and spilling along a shallow sill of the sound towards the ocean (Okely et al. 2007). The experiment showed that the dye had dispersed beyond what could be visually detected within a distance of around 0.9 miles (1.5 kilometers), well within the protected deeper region of Cockburn Sound about three miles (five kilometers) from the diffuser. The environmentally benign dye experiment was first commissioned in December 2006 and repeated in April 2007 when conditions were calm.

The key challenges associated with selecting the most appropriate location for the desalination plant's ocean outfall discharge are: finding an area void of endangered species and stressed marine habitats; identifying a location with strong ocean currents that allows quick and effective dissipation of the concentrate discharge; avoiding areas with busy naval vessel traffic, which could damage the outfall facility and change mixing patterns; and selecting a discharge location in relatively shallow waters that are close to the shoreline to minimize outfall construction expenditures. Some key advantages of constructing a new ocean outfall are that this type of concentrate disposal option can accommodate practically any size of seawater desalination plant and provides for more freedom in selecting a plant location, as compared to the other two open-ocean disposal approaches where existing outfalls are used.

Courtesy of Water Corporation of Western Australia

Figure 4-8 Perth desalination plant discharge diffuser – rhodamine dye test

Key disadvantages of this discharge approach are that it is usually very costly and that its implementation requires extensive environmental and engineering studies. Depending on the site-specific conditions, the costs for a new ocean outfall could be significant, and may range from 5 to 30 percent of the total desalination plant construction expenditures. The higher end of this range tends to apply to large desalination plants (i.e., facilities of freshwater production capacity of 10 mgd [38 MLD] or more) with long outfall pipelines, where the construction of a new concentrate outfall may represent more than 20 percent of the total facility cost.

Discharge Through Existing Wastewater Treatment Plant Outfall—environmental impacts and mitigation measures.

Discharging concentrate through an existing wastewater treatment plant outfall would generally involve direct disposal through the plant outfall. While it would also be possible to discharge into a nearby wastewater collection system, thereby indirectly utilizing the wastewater outfall, this approach would be expected to have a considerable impact on the wastewater treatment facility operation as well as corrosion concerns within the collection system. The key feature of the combined discharge method is the accelerated mixing due to the blending of heavier than ocean water concentrate with the lighter wastewater discharge. Depending on the volume of the concentrate and on how well the two waste streams are mixed prior to the point of discharge, the blending may reduce the size of the wastewater discharge plume and dilute some of its constituents. This co-discharge with the lighter-than-seawater wastewater effluent would also accelerate the dissipation of the saline plume by floating this plume upward and expanding the volume of the ocean water with which it mixes.

Direct discharge through an existing wastewater treatment plant outfall has found a limited application to date, especially for medium and large seawater desalination plants. This disposal method had been practiced during the short-lived operations of the Santa Barbara seawater desalination plant in California (Figure 4-9). There, the desalination plant concentrate discharge volume was comparable to that of the wastewater treatment plant effluent discharge (5.5 mgd or 21 MLD).

Courtesy of Water Globe Consulting

Figure 4-9 5.5 MGD Santa Barbara seawater desalination plant, California

One key consideration related to the use of an existing wastewater treatment plant outfall for direct seawater desalination plant concentrate discharge is the availability and cost of wastewater outfall capacity. For this concentrate disposal option to be feasible, an existing wastewater treatment plant in the vicinity of the desalination plant must exist, and this plant must have available outfall discharge capacity. Additionally, the fees associated with the use of the wastewater treatment plant outfall must be reasonable.

Another key consideration is the potential for WET of the blended discharge that may result from ion imbalance of the blend of the two waste streams. The wastewater utility involved must be comfortable with the handling and separation of liability for environmental impacts due to the blended discharge between the owner of the desalination plant and the owner of the wastewater treatment plant. Rarely is this beneficial combination of conditions easy to find, especially when discharging large seawater concentrate volumes.

In an example of blending increasing WET, bioassay tests completed on blends of desalination plant concentrate and wastewater effluent from the El Estero Wastewater Treatment Plant in Santa Barbara, California indicate that the blend could exhibit toxicity on fertilized sea urchin *(Strongylocentrotus purpuratus)* eggs. Parallel tests on desalination plant concentrate diluted to similar TDS concentration with seawater rather than wastewater effluent did not show such toxicity effects on sea urchins. Similarly, long-term exposure of red sea urchins to the blend of concentrate from the Carlsbad seawater desalination demonstration plant and ambient seawater discharged by the adjacent Encina Power Plant confirms that sea urchins can survive elevated salinity conditions when the discharge is void of wastewater.

The most likely factor causing the toxicity effect on the sensitive marine species is the difference in ratios between the major ions (Ca, Mg, Na, Cl, and SO_4) and TDS that occur in the wastewater effluent-concentrate blend as compared to the concentrate-seawater blend and the ambient ocean water. Because the SWRO membranes reject all key seawater ions at approximately the same level, the ratios between the concentrations of the individual key ions that contribute to the seawater salinity (Ca, Mg, Na, Cl, and SO_4) and the TDS in the concentrate are approximately the same as the ratios in the ambient seawater. Depending on the individual source of wastewater, the ionic ratio for the wastewater likely varies somewhat from ambient seawater, causing the toxicity effect.

Another important issue to be considered when an existing wastewater discharge with diffusers is used for codisposal of wastewater and seawater concentrate is the fact that the buoyancy of the mix will be reduced, and the wastewater discharge may no longer provide adequate mixing unless the diffusers are reconfigured. The need for diffuser reconfiguration to accommodate mixed wastewater effluent/concentrate discharge can be established by hydrodynamic modeling.

Diffuser reconfiguration may involve closing (capping) of some of the existing diffusers, modifying the discharge diffuser structure, or pumping of the mix of concentrate and wastewater influent in order to increase the kinetic energy of the discharge that is available for mixing with the ambient seawater.

Use of existing wastewater treatment plant outfalls for concentrate discharge also has some key advantages including avoiding costs and environmental impacts associated with the construction of a new outfall for the seawater desalination plant. Mixing of the positively buoyant wastewater discharge with the negatively buoyant concentrate promotes accelerated dissipation of both the wastewater plume and the concentrate. In addition, concentrate often contains metals, organics, and pathogens at significantly lower levels than the wastewater discharge, which reduces the overall waste discharge concentration of the mix.

Although the use of existing wastewater treatment plant outfalls or concentrate discharge to the sanitary sewer system may seem attractive for its simplicity and low construction costs, this disposal method has a number of limitations. Because of the potential toxicity effects of the concentrate-wastewater effluent blend, the direct discharge of the seawater concentrate through existing wastewater discharge outfalls may be limited to relatively small concentrate discharge flows. Similarly, indirect discharge of the concentrate through the wastewater collection system may be severely constrained or practically impossible especially if the wastewater plant effluent is reused for irrigation.

Discharge Through Existing Power Plant Outfall (Colocation)—environmental impacts and mitigation measures.

The key feature of the colocation concept is the direct connection of the membrane desalination plant intake and discharge facilities to the discharge outfall of an adjacently located coastal power generation plant. This approach allows the use of the power plant cooling water as both the source water for the seawater desalination plant and as a blending water to reduce the salinity of the desalination plant concentrate prior to the discharge to the ocean. Figure 4-10 illustrates the conceptual implementation of the colocation approach for a 50 mgd (189 MLD) seawater desalination plant planned for Carlsbad, California.

As shown on Figure 4-10, under typical operational conditions, approximately 600 mgd (2300 MLD) of seawater enters the power plant intake facilities and is screened and pumped through the plant's condensers for cooling. The cooling water discharged from the condensers is typically 5 to 10°C warmer than the ambient ocean water and is conveyed to the ocean via a separate discharge canal. The Carlsbad desalination plant intake structure will be connected to the end of this discharge canal, and under normal operational conditions, the intake would divert approximately 100 mgd of the 600 mgd (380 to 2300 MLD) of cooling water for desalination. The desalination would yield approximately 50 mgd (190 MLD) of fresh water for potable use. The remaining 50 mgd (190 MLD) will have salinity approximately two times that of the ocean water (i.e., 67,000 mg/L). This seawater concentrate will be returned to the power plant discharge canal for blending with the cooling water prior to discharge to the Pacific Ocean. Under average conditions, the blend of 500 mgd (1900 MLD) of cooling water and 50 mgd (190 MLD) of concentrate would have discharge salinity of 36,200 mg/L, which is within the natural fluctuation of the ocean water salinity in the vicinity of the existing power plant discharge.

66 DESALINATION OF SEAWATER

Colocation of large scale desalination with a power station was first utilized in the United States for the Tampa Bay Seawater Desalination Project, and since then has been considered for numerous plants in the United States and worldwide. The intake and discharge of the Tampa Bay Seawater Desalination Plant are connected directly to the cooling water discharge outfalls of the Tampa Electric (TECO) Big Bend Power Station (Figure 4-11).

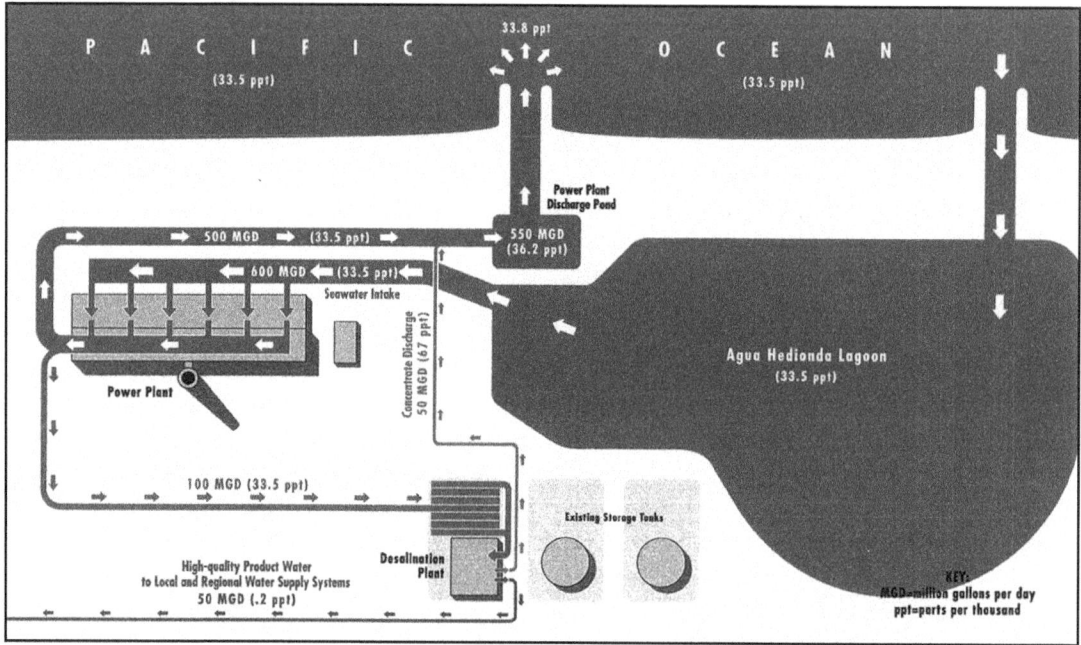

Courtesy of Water Globe Consulting

Figure 4-10   Colocation concept for the Carlsbad Seawater Desalination Plant

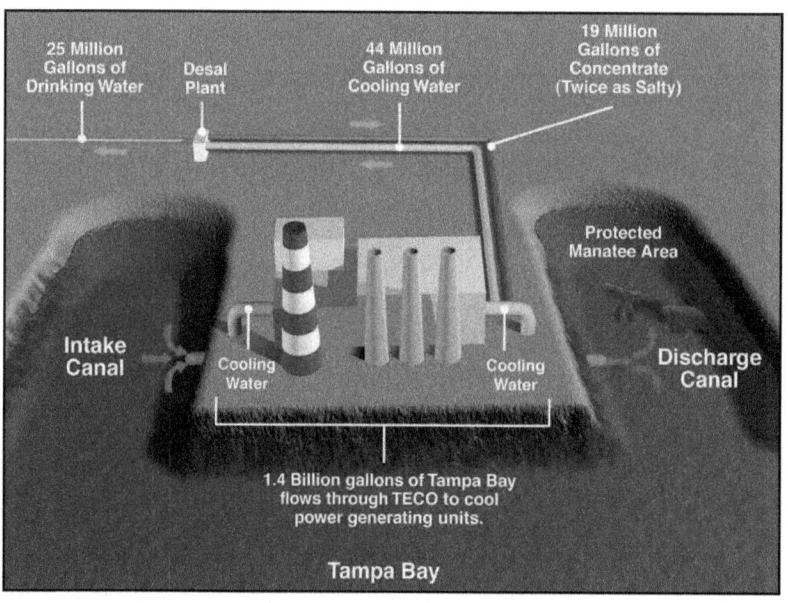

Courtesy of Water Globe Consulting

Figure 4-11   Colocation of Tampa Bay Seawater Desalination Plant

An example of a colocation configuration where the power plant discharge is used only for dilution of the concentrate and not as a source of feedwater is the 32 mgd (121 MLD) Carboneras desalination plant in Spain (Figure 4-12), which currently is one of the largest SWRO plants in Europe. The concentrate is discharged to the cooling water canal of a nearby coastal power generation plant and diluted to environmentally safe levels before returning to the sea. The Carboneras seawater desalination plant has a separate open intake independent from the power plant.

Colocation with an existing once-through cooling coastal power plant yields four key benefits: (1) the construction of a separate desalination plant outfall structure is avoided thereby reducing the overall cost of desalinated water; (2) the salinity of the desalination plant discharge is reduced as a result of the mixing and dilution of the membrane concentrate with the power plant discharge, which has ambient seawater salinity; (3) because a portion of the discharge water is converted into potable water, the power plant thermal discharge load is decreased, which in turn lessens the negative effect of the power plant thermal plume on the aquatic environment; and (4) the blending of the desalination plant and the power plant discharges results in accelerated dissipation of both the salinity and the thermal discharges.

The cost of construction of a separate ocean outfall could be significant, and its avoidance would result in a measurable reduction of plant construction expenditures. In addition, the length and configuration of the desalination plant concentrate discharge outfall are closely related to the discharge salinity. Usually, the lower the discharge salinity, the shorter the outfall and the less sophisticated the discharge diffuser configuration needed to achieve environmentally safe concentrate discharge. Blending the desalination plant concentrate with the lower salinity power plant cooling water reduces the overall salinity of the ocean discharge within the range of natural variability of the ocean, thereby completely eliminating the need for complex and costly discharge diffuser structures.

Courtesy of Water Globe Consulting

Figure 4-12 32 MGD Carboneras SWRO plant in Spain

Deep Well Injection and Coastal Well Disposal—environmental impacts and mitigation measures.

Deep well injection involves the discharge of desalination plant concentrate into an acceptable confined deep underground aquifer below the freshwater aquifer(s) using a system of disposal wells. The deep-well injection concentrate disposal system also includes a set of monitoring wells to confirm that the concentrate is not migrating to the adjacent aquifers. A variation of this disposal alternative is the injection of concentrate into existing oil and gas fields to aid field recovery.

Deep well injection is frequently used for concentrate disposal from all sizes of brackish water desalination plants. Beach well disposal is an alternative concentrate disposal practice.

Unlike deep well injection, beach well disposal consists of concentrate discharge into a relatively shallow unconfined coastal aquifer that ultimately conveys the discharge into the open ocean through the ocean bottom. Beach wells are used for small- and medium-sized seawater desalination plants and are not discussed further in this manual due to limited application and success record.

Deep well injection is most widely used for brackish water discharge. This method of concentrate disposal has found very limited application for seawater desalination. Key considerations associated with this concentrate disposal method include:

- Limited to site-specific conditions of confined aquifers of large storage capacity that have good soil transmissivity (Swartz 2000).

- Not feasible for areas of elevated seismic activity or near geologic faults that can provide a direct hydraulic connection between the discharge aquifer and a water supply aquifer.

- Potential for contamination of groundwater with concentrated pollutants, if the discharge aquifer is not adequately separated from the water supply aquifer in the area of discharge.

- Potential leakage from the wells.

- Potential scaling and decrease of well discharge capacity over time.

- A backup concentrate disposal method is required for periods of time when the injection wells are tested and maintained.

- High well construction and monitoring costs.

The cities of Marina and Sand City, California have used shallow coastal well discharge of seawater desalination plant concentrate. The discharges inject concentrate with salinity between 30,000 and 43,000 mg/L of TDS into shallow dune sand aquifers via a conventional well for Marina and a horizontal well for Sand City. The injected concentrate blends with groundwater and ultimately diffuses into the turbulent surf zone. At present, the concentrate from the Marina SWRO plant is no longer discharged through the beach well because of severe scaling problems. Also, the Marina Coast Water District is planning on building a new larger seawater desalination plant that will codispose the seawater concentrate with the wastewater effluent from a nearby wastewater treatment plant. The 0.6 mgd (2.3 MLD) Sand City desalination facility began operation in 2009 and continues to discharge through its subsurface discharge. This plant uses intake wells to collect source seawater (typically of salinity of 17,000 and 28,000 mg/L) for the production of freshwater and collects additional seawater to dilute the desalination plant concentrate to ambient seawater salinity levels as needed (Wetterau et al. 2009).

A recent study in Spain suggests that actual dilution of the concentrate from a beach-discharge outfall may be lower than normally accepted (Fernandez-Torquemada et al. 2005). In this case, elevated salinity was reported in deep localities several miles from the discharge point. Similarly, a modeling study from Oman suggests that continuously discharging concentrate directly on the shoreline may result in increased salinity along the coastline (Purnama et al. 2003).

Beneficial Use of Concentrate

Concentrate from seawater desalination plants contains large quantities of minerals that may have commercial value when extracted. The most valuable minerals from the seawater are magnesium, calcium, sodium, chloride, and bromide. Magnesium compounds in seawater have agricultural, nutritional, chemical, construction, and industrial applications. Calcium sulphate (gypsum) is used as construction material for wallboard, plaster, building cement, and road building and repair. Sodium chloride can be used for production of chlorine and caustic soda, highway de-icing, and food products. Technologies for beneficial recovery of minerals from concentrate can be used for management of concentrate from both inland brackish water desalination plants and coastal seawater desalination plants. These technologies have the potential to decrease the volume and cost of transporting concentrate as well.

The existing salt recovery technologies extract salts by fractional crystallization or precipitation. Crystallization of a given salt can be achieved by concentrate evaporation or temperature control. Fractional precipitation is attained by adding a chemical precipitating agent to selectively remove a target mineral from the concentrate solution. For example, there is a commercially available technology for extraction of magnesium and calcium salts from concentrate and for production of structural materials from these salts.

MANAGEMENT OF DESALINATION PLANT RESIDUALS

The key residuals generated at a desalination plant, apart from the concentrate discussed previously, include pretreatment process waste streams and spent membrane cleaning solutions. The most commonly used methods for management of these residuals are summarized in Table 4-3.

Management of Pretreatment Process Residuals

Table 4-4 presents types of residuals that may be produced in the pretreatment process before applying membrane desalination. The amount of residuals produced is primarily a function of the feedwater quality relative to the constituents that must be removed prior to the membrane desalination process. Seawater from open ocean intakes contains significant levels of suspended solids. These solids must be removed prior to reverse osmosis, either in a backwash stream or as sludge. Other than the concentrate stream, these solids create the most significant residual stream from a desalination plant.

Spent filter backwash water is a waste stream produced by the membrane plant's pretreatment filtration system. Depending on the type of pretreatment system used (granular or membrane filters), the spent filter backwash water will vary in quantity and quality. In general, the membrane pretreatment systems produce 1.5 to 2 times larger volume of spent filter backwash water than the granular media filters. However, compared to microfiltration (MF) or ultrafiltration (UF) membrane pretreatment filters, granular media filters typically require larger dosages of coagulant for pretreatment. Depending on the source water quality, membrane use may allow successful pretreatment without coagulant addition. Spent pretreatment filter backwash water may include filter aids and coagulants.

Discharge to the ocean, along with plant concentrate without treatment, is one of the most widely practiced disposal methods for spent filter backwash water internationally. This is typically the lowest cost disposal method because it does not involve any treatment prior to disposal. However, the practice is less common in more environmentally sensitive areas, such as Australia and the U.S.

Table 4-3   Residuals from seawater desalination processes

Residual	Source or Cause	Most Common Disposal Methods
Filter and screening backwash solids/sludge	Suspended solids in the feedwater. May contain coagulants and/or filter aid polymers.	Settling/thickening, dewatering by belt filter presses and sludge disposal to sanitary landfill.
Backwash water	From removal of suspended solids in the feed water.	Recycle to the pretreatment filtration system for reuse or disposal with concentrate.
Cleaning solutions	Cleaning of filtration membranes (MF/UF) and RO membranes.	Blending with concentrate and disposal to surface water body. Disposal to sanitary sewer.
Spent media (sand, anthracite and/or garnet)	From the removal of suspended solids in the source water.	Disposal to sanitary landfill.
Cartridge Filters – polypropylene	Final fine filtration prior to RO, periodic replacement.	Disposal to sanitary landfill.
MF/UF pretreatment membranes – polymeric material (polypropylene, polysulphone, polyvinylidenefluroide [PVDF], cellulose acetate.	Membrane replacement for MF/UF pretreatment systems.	Disposal to sanitary landfill. Regeneration by membrane manufacturer and reuse.
RO membranes (polyamide thin film composite, cellulose acetate)	Membrane replacements.	Disposal to sanitary landfill. Regeneration by membrane manufacturer and reuse.

Courtesy of Water Globe Consulting

On-site treatment prior to surface water discharge or recycle upstream of the filtration system is another option. The filter backwash water must be treated when its direct discharge does not meet water quality requirements. Typically, the most widely used granular media backwash treatment method is gravity settling in lamella plate sedimentation tanks.

Spent wash water from membrane pretreatment systems is less likely to contain coagulants than media filtration backwash, making it more amenable to discharge with the RO concentrate. Where membrane filtration backwash is recovered, it will employ similar processes as are used with media filters. The solid residuals (sludge) retained in the sedimentation basin are generally dewatered onsite in a designated solids handling facility.

Conventional granular media filters and membrane pretreatment systems differ significantly by the type, quality, and quantity of the residuals generated during the filtration process (see Table 4-4).

Typically, granular media filtration systems generate only one large liquid waste stream, containing both spent filter backwash and filter-to-waste flows. The volume of this stream in a well designed plant varies between 2 to 6 percent of the total plant intake seawater volume. In addition to the particulate solids and colloids that are contained in the source seawater, this waste stream also contains coagulant (typically iron salt) and may have flocculant (polymer).

Membrane pretreatment systems generate two primary liquid residual streams: (1) spent membrane backwash water (reject) and (2) membrane cleaning solution from daily or periodic chemically enhanced backwash (CEB). The volume of the spent membrane filter backwash water is typically 5 to 10 percent of the plant intake source volume – i.e., approximately one-and a half to two times larger than the spent filter backwash water volume of granular media pretreatment systems.

The difference in total liquid residual volume generated by membrane pretreatment systems is even larger, taking into account that the microscreens needed to protect the membrane pretreatment filters will be a source of an additional waste discharge from their intermittent cleaning. While conventional traveling fine bar screens use 0.1 to 0.2 percent of the intake source water for cleaning, the microscreens generate waste screen-wash volumes that equal a slightly higher to 0.1 to 0.5 percent of the intake flow. The larger waste stream volume of the membrane pretreatment system would require proportionally larger intake seawater volume, which in turn would result in increased size and construction costs for the desalination plant intake facilities, and pump station, and in higher O&M costs for source seawater pumping to the pretreatment facilities.

In addition to backwashing, cost competitive design and operation of membrane pretreatment systems often require daily or periodic chemically enhanced membrane backwash (CEB) using a large dosage of chlorine (typically 20 to 200 mg/L) and strong base and/or acid over a short period of time. This performance-enhancing CEB adds to the volume of the waste streams generated at the RO membrane plant and to the overall cost of source water pretreatment. The daily volume of waste stream generated during CEB is usually 0.2 to 0.4 percent of the volume of the intake source seawater.

Another waste stream that is associated only with membrane pretreatment is generated during the periodic chemical cleaning of the pretreatment membranes. Extended off-line chemical cleaning, often referred to as clean-in-place (CIP), is usually needed once every one to three months. During CIP, membranes are soaked in a solution of hydrochloric or citric acid, sodium hydroxide, sodium hypochlorite, and/or surfactants. CIP is critical for maintaining steady state membrane performance and productivity, but the cleaning generates an additional waste stream that is 0.03 to 0.05 percent of the source seawater volume.

One key advantage of membrane pretreatment systems is that the waste filter backwash generated by these systems contains less source water conditioning chemicals (coagulant and polymer), and therefore it is more environmentally benign compared to the waste filter backwash stream generated by conventional pretreatment facilities. This benefit stems from the fact that typically coagulant dosage for seawater pretreatment by membrane filtration, if used at all, is two to three times lower than that for granular media filtration.

In many cases, source seawater may not need to be conditioned with coagulant before membrane pretreatment, and this spent filter backwash could be disposed of along with the SWRO concentrate without further treatment. Conversely, due to the high content of iron, the spent filter backwash from granular media filtration pretreatment would need to be treated by sedimentation and other processes, with the settled solids dewatered and disposed to sanitary landfill; otherwise, the high content of iron salt in the backwash water will cause the desalination plant discharge to have a red color, as has been the case for some seawater outfalls.

Table 4-4 Comparison of waste streams from granular media and membrane pretreatment

Waste Stream	Granular Media Filtration (% of Feed Volume)	Membrane Filtration (% of Feed Volume)
Intake Bar Screens Wash-water	0.1 – 0.2	0.1 – 0.2
Microscreen Wash-water	None (Not Needed)	0.1 – 0.5
Spent Filter Backwash Water (Reject)	2.0 – 6.0	5.0 – 10.0
Chemically Enhanced Backwash	None (Not Needed)	0.2 – 0.4
Spent Membrane Cleaning Chemicals	None (Not Needed)	0.03 – 0.05
Total Waste (% of Feed Volume)	2.1 – 6.2	5.4 – 11.1

Courtesy of Water Globe Consulting

The waste streams generated during the CEB and the CIP membrane cleaning should be pretreated on-site in a neutralization tank, prior to discharge. The additional treatment and disposal costs of the waste membrane cleaning chemicals should be considered when comparing membrane and granular media pretreatment systems.

GREENHOUSE GAS EMISSIONS—IMPACTS AND MANAGEMENT

Gases that trap heat in the atmosphere are referred to as greenhouse gases (GHG) (USEPA 2006). Some greenhouse gases such as carbon dioxide occur naturally and are emitted to the atmosphere through natural processes and human activities. Other greenhouse gases (e.g., fluorinated gases) are created and emitted primarily through human activities. The principal greenhouse gases that enter the atmosphere because of human activities are carbon dioxide, methane, nitrous oxide, and fluorinated gases.

Carbon Dioxide (CO_2): Carbon dioxide enters the atmosphere through the burning of fossil fuels (oil, natural gas, and coal), solid waste, trees and wood products, and also as a result of other chemical reactions (e.g., manufacture of cement). Carbon dioxide is also removed from the atmosphere (or *sequestered*) when it is absorbed by plants as part of the biological carbon cycle.

Methane (CH_4): Methane is emitted during the production and transport of coal, natural gas, and oil. Methane emissions also result from livestock and other agricultural practices and by the decay of organic waste in municipal solid waste landfills.

Nitrous Oxide (N_2O): Nitrous oxide is emitted during agricultural and industrial activities, as well as during combustion of fossil fuels and solid waste.

Fluorinated Gases: Hydrofluorocarbons, perfluorocarbons, and sulfur hexafluoride are synthetic, powerful greenhouse gases that are emitted from a variety of industrial processes. These gases are typically emitted in smaller quantities, but because they are potent greenhouse gases, they are sometimes referred to as High Global Warming Potential gases (High GWP gases).

Changes in the atmospheric concentrations of these greenhouse gases can alter the balance of energy transfers between the atmosphere, space, land, and the oceans and ultimately result in global and local climate variability and permanent changes (NRC 2001). Many elements of human society and the environment are sensitive to climate variability and change. Human health, agriculture, natural ecosystems, coastal areas, and heating and cooling requirements are examples of climate-sensitive systems. The extent of climate change effects, and whether these effects prove harmful or beneficial, will vary by region, over time, and with the ability of different societal and environmental systems to adapt to or cope with the change.

Rising average temperatures are already affecting the environment. Some observed changes include shrinking of glaciers, thawing of permafrost, later freezing and earlier break-up of ice on rivers and lakes, lengthening of growing seasons, shifts in plant and animal ranges and earlier flowering of trees (IPCC 2007).

Global temperatures are expected to continue to rise as human activities continue to add carbon dioxide, methane, nitrous oxide, and other greenhouse (or heat-trapping) gases to the atmosphere. Most of the United States is expected to experience an increase in average temperature as a result of increase in greenhouse gas emissions (IPCC 2007).

According to a recent USEPA GHG emission inventory, the primary greenhouse gas emitted by human activities in the United States in 2006 was carbon dioxide, representing approximately 84.8 percent of total greenhouse gas emissions (USEPA 2008). The largest source of carbon dioxide and of overall greenhouse gas emissions is fossil-fuel based production of electricity. The second largest source is transportation. Despite the attention of some environmental groups to greenhouse emissions associated with water production, both conventional and membrane water treatment plants are typically not major sources of GHGs.

For example, in California, where approximately 19 percent of the total electricity use is employed to treat and transport water (the highest of any state), only 1 percent of this energy is related to the actual treatment of the water.

Greenhouse Gas Emission Management

Management of GHG associated with the operation of water treatment plants is a relatively new practice in the United States. In Australia, GHG considerations have played major roles in the implementation of seawater desalination facilities, with desalination plants in Perth, Queensland, Sydney, Adelaide, and Victoria all employing wind power as their primary energy source. In the United States, solar power is planned as the primary energy source for the Cambria desalination plant, currently under development in California, while other desalination projects are evaluating renewable energy to supply a portion of the total energy needs. The key step in GHG management is the development of a Climate Action Plan (CAP), which defines the carbon footprint of desalination plant operations as well as identifies a portfolio of alternative technologies and measures to achieve carbon footprint project neutrality: from the use of state-of-the-art energy reduction technologies to the implementation of renewable energy projects, and of carbon dioxide sequestration initiatives including on-site carbon dioxide use, reforestation, and coastal wetland restoration.

An example of the key steps and approaches for the development of CAP is presented in a case study for the 50 mgd (189 MLD) Carlsbad seawater desalination plant. As indicated previously, this project is collocated with the Encina coastal power generation station, which currently uses seawater for once-through cooling (Figure 4-13).

The following are key components of the Climate Action Plan.

Assessing project gross carbon footprint. The carbon footprint of the seawater desalination plant is the amount of greenhouse gases that would be released into the air from the power generation sources that will supply electricity for the plant. Usually, carbon footprint is measured in pounds (lb) or metric tons of carbon dioxide (CO_2) emitted

Courtesy of Water Globe Consulting

Figure 4-13 Carlsbad seawater desalination project

per year. The total plant carbon footprint is dependent on two key factors: (1) how much electricity is used by the desalination plant; and (2) what sources (fossil fuels, wind, sunlight, etc.) are used to generate the electricity supplied to the plant. Both of these factors could be variable over time and therefore, the Climate Action Plan has to have the flexibility to incorporate such changes.

The Carlsbad seawater desalination plant is planned to be operated continuously, 24 hours a day/ 365 days per year, and to produce an average annual drinking water flow of 50 mgd (189 MLD). When the plant was originally conceived, the total baseline power use for this plant was projected at 31.3 megawatts (MW) or 15.03 kWh/1,000 gal (3.97 kWh/m^3) of drinking water. This power use incorporates both production of fresh drinking water and conveyance, and delivery of this water to the distribution systems of the individual utilities and municipalities served by the plant.

However, over the lengthy period of project permitting, the seawater desalination technology has evolved. By taking advantage of the most recently available state-of-the art technology for energy recovery and by advancing the design to accommodate latest high efficiency RO system feed pumps and membranes, the actual project power use was reduced to 13.48 kWh/1,000 gal (3.6 kWh/m^3) of drinking water. As a result, the total annual energy consumption for the Carlsbad seawater desalination project used to determine the plant carbon footprint is 246,000 MWh/yr. This energy use is determined for an annual average plant production capacity of 50 mgd (189 MLD). As actual production capacity may vary from year to year, so would the total energy use.

To determine the carbon footprint (CF) from the desalination plant annual energy use, the demand is multiplied by the electric grid emission factor (Emission Factor), which is the amount of greenhouse gases emitted during the production of unit electricity consumed from the power transmission and distribution system:

CF (lb of CO_2/yr) = Annual Plant Electricity Use (MWh/yr) × Emission Factor (lb of CO_2/MWh)

The actual value of the Emission Factor is specific to the actual supplier of electricity for the project. The San Diego Gas and Electric (SDG&E) supplies electricity to the Carlsbad seawater desalination project.. Similar to other power suppliers in California, SDG&E determines their Emission Factor based on a standard protocol developed by the California Climate Action Registry (CCAR). CCAR was created by California Legislature (SB 1771) in 2001 as a nonprofit voluntary registry for GHG emissions and is the authority in California that sets the rules by which GHG emissions are determined and accounted for.

Based on information provided in their most recent emissions report (CCAR 2008), the SDG&E emission factor is 546.46 lb (247.96 kg) of CO_2 per MWh of delivered electricity. At 246,000 MWh/yr of energy use and 546.46 lb (247.96 kg) CO_2/MWh, the total carbon footprint for the Carlsbad seawater desalination project is calculated at 134.4 million lb of CO_2 per year (61,100 metric tons CO_2/yr). This carbon footprint is reflective of the latest energy efficient design of the desalination plant. A more conventional desalination plant design (274,000 MWh/hr) would have a carbon footprint of 68,100 metric tons CO_2/yr.

Offsetting carbon footprint by reduced water imports. In many parts of the world such as Spain, Israel, Singapore, Australia, and California, seawater desalination plants are built to replace in- or out-of-state water transfers/supplies. Long-distance water transfers are often very energy intensive and the carbon footprint of such water supply alternatives may be comparable to that of desalination plant of similar capacity. Offsetting the carbon footprint of such long-distance water transfers by building local desalination plants can be counted as a carbon-footprint reduction measure for the desalination plant.

For example, San Diego County currently imports approximately 80 percent of its water from two sources – the Sacramento Bay/San Joaquin River Delta, traditionally known as the Bay-Delta, and the Colorado River. The imported Bay-Delta water is withdrawn from the source and conveyed via a complex system of intakes, dams, reservoirs, aqueducts, and pump stations (State Water Project), and treated in conventional water treatment plants prior to its introduction to the water distribution system. The total amount of electricity needed to deliver this water to San Diego County via the State Water Project facilities is 10.45 kWh/1,000 gal (2.76 kWh/m^3), which includes 9.93 kWh/1,000 gal (2.62 kWh/m^3) for delivery, 0.21 kWh/1,000 gal (0.06 kWh/m^3) for evaporation losses, and 0.31 KWh/1,000 gal for treatment.

Over the past decade, the availability of imported water from the State Water Project has been in a steady decline due to prolonged drought, climate change patterns, and environmental and population growth pressures. One of the key reasons for the development of the Carlsbad seawater desalination project is to replace 50 mgd (189 MLD) of the water imported via the State Water Project with fresh drinking water produced locally by tapping the ocean as an alternative drought-proof source of water supply. Because the desalination project will offset the import of 50 mgd (189 MLD) of water via the State Water Project, once in operation, this project will also offset the electricity consumption of 10.45 kWh/1,000 gal (2.76 kWh/m^3), and the GHG emissions associated with pumping, treatment, and distribution of this imported water. The annual energy use for importing 50 mgd (189 MLD) of State Water Project water is therefore 190,700 MWh/yr. At 546.46 lb (247.96 kg) CO_2/MWh, the total carbon footprint of the water imports that will be offset by desalinated water is therefore 104.2 million lbs of CO_2 per year (47,400 metric tons CO_2/yr).

Considering that the gross carbon footprint of the desalination plant is 61,100 metric tons CO_2/yr, and that 47,400 metric tons CO_2/yr (77.4 percent) of these GHG emissions would be offset by reduction of 50 MGD (189 MLD) of water imports to San Diego County, the Carlsbad desalination plant's net carbon footprint is estimated at 13,700 metric tons CO_2/yr.

Climate Action Plan for net carbon footprint reduction. The main purpose of the Climate Action Plan for a given seawater desalination project is to eliminate a plant's net carbon footprint by implementing measures for: energy efficient facility design and operations; green building design; use of carbon dioxide for water production; on-site solar power generation; carbon dioxide sequestration by creation of coastal wetlands and reforestation; funding renewable power generation projects; and acquisition of renewable energy credits. Project carbon neutrality would be achieved by a balanced combination of these measures.

The size and priority of the individual projects included in the Climate Action Plan should be determined based on a life-cycle cost-benefit analysis and overall benefit for the local community. Implementation of energy efficiency measures for water production, green building design, and carbon dioxide sequestration projects in the vicinity of the project site should be given the highest priority.

The project Climate Action Plan is a living document that must be updated periodically in order to reflect the dynamics of development of desalination and green energy generation technologies; and the efficiency and cost-effectiveness of various carbon footprint reduction and offset alternatives must also be updated.

Green building design. Whenever practical and viable, the desalination plant should be located on a site of little current value or public use. Reclaiming low-value land will reduce project imprint on the environment as compared to using a new undisturbed site. For example, the Carlsbad seawater desalination plant will be located on a site occupied by dilapidated fuel oil tanks. The tanks and their contents will be removed and the site will be reclaimed and reused to construct the desalination plant.

Another approach to reduce a desalination plant's physical imprint on the environment is to minimize desalination plant site footprint. For example, a key green feature of the Carlsbad seawater desalination plant design is its compactness. The desalination plant facilities will be configured as series of structures sharing common walls, roofs, and equipment, which will allow significant reduction of its physical footprint. The total area occupied by the desalination plant facilities will be less than 7 acres (3 hectares). When built, this will be the smallest footprint desalination plant in the world per unit production capacity (7 acres per 50 mgd or 3 hectares per 190 MLD). For comparison, the 25 mgd (95 MLD) Tampa Bay seawater desalination plant occupies 8 acres (3.2 hectares); the 70 mgd (265 MLD) Orange County Groundwater Recharge Project, which also uses a reverse osmosis system, occupies approximately 40 acres (16 hectares); and the 86 mgd (325 MLD) Ashkelon, Israel seawater desalination plant, which currently is the largest operational seawater RO facility in the world, occupies 24 acres (9.7 hectares). A plant with a smaller physical footprint will also yield a smaller construction-related carbon footprint resulting in lower construction material expenditures and GHG emissions from construction equipment due to smaller volume of excavation and concrete works. Reduced construction site footprint also generates less dust emissions and requires less water for dust control.

Whenever economically viable and practical, building design should follow the principles of the Leadership in Energy and Environmental Design (LEED) program. This is a program of the U.S. Green Building Council and was developed to promote construction of sustainable buildings that reduce the overall impact of building construction and functions on the environment by (1) sustainable site selection and development; (2) energy efficiency; (3) materials selection; (4) indoor environmental quality; and (5) water savings.

Consistent with the principles of the LEED program, the desalination plant buildings should include features and materials that allow minimizing energy use for lighting, air conditioning, and ventilation. For example, a portion of the walls of the main desalination plant building of the Carlsbad seawater desalination plant will be equipped with translucent panels to maximize daylight use and views to the outside. Nonemergency interior lighting will be automatically controlled to turn off in unoccupied rooms and facilities. A monitoring system will ensure that the ventilation in the individual working areas in the building is maintained at its design minimum requirements. In addition, building design will incorporate water conserving fixtures (lavatory faucets, showers, water closets, urinals, etc.) for plant staff service facilities and for landscape irrigation.

The green desalination plant buildings should incorporate low-emitting materials and thus pose less risk to the natural environment and building's occupants. Low emitting paints, coatings, adhesives, sealants and carpet systems should be used on the interior of the buildings whenever possible. The building design team should include professional engineers that have achieved the LEED Accredited Professional designation and are well experienced with the design and construction of green buildings.

Use of carbon dioxide for water production. Approximately 2,100 metric tons of carbon dioxide per year are planned to be used at the desalination plant for post-treatment of the freshwater (permeate) produced by the RO system. Carbon dioxide in a gaseous form will be added to the RO permeate in combination with calcium hydroxide or calcium carbonate to form soluble calcium bicarbonate, which adds hardness and alkalinity to the drinking water for distribution system corrosion protection. In this posttreatment process of RO permeate stabilization, gaseous carbon dioxide is sequestered into soluble form of calcium bicarbonate. Because the pH of the drinking water distributed for potable use is in a range of 8.3 to 8.5 when CO_2 is in a soluble bicarbonate form, the carbon dioxide introduced in the RO permeate would remain permanently sequestered in this form and ultimately would be consumed with the drinking water. The plant designer/owner should require the supplier of carbon dioxide to guarantee that the source of this gas is a waste recovery process (i.e., the gas is generated as a waste sidestream, which if not captured and used, will be released into the atmosphere). This requirement is not very difficult to comply with because most of the commercial suppliers of carbon dioxide use gas generated as a waste from other industrial processes (i.e., ethanol plants, breweries, etc.), which is then purified and sold as a commercial product.

Carbon dioxide sequestration in coastal wetlands. In addition to the benefit of marine habitat restoration and enhancement, coastal wetlands also act as a "sink" for carbon dioxide. Tidal wetlands are very productive habitats that remove significant amounts of carbon dioxide from the atmosphere, a large portion of which is stored in the wetland soils. While freshwater wetlands also sequester carbon dioxide, they are often a measurable source of methane emissions. For comparison, coastal wetlands and salt marshes release negligible amounts of greenhouse gases, and therefore, their carbon sequestration capacity is not measurably reduced by methane production.

For example, as a part of the Carlsbad seawater desalination project, Poseidon Resources is planning to develop 37 to 68 acres (15 to 28 hectares) of new coastal wetlands in San Diego County. These wetlands will be designed to create habitat for marine species similar to those found in the Agua Hedionda Lagoon (see Figure 4-13), from which source seawater is collected for the power plant and for desalination plant operations. Once the wetlands are fully developed, they will be maintained and monitored over the life of the desalination plant operations. Site-specific research is planned to quantify the actual carbon sequestration capacity of the new wetland system proposed development as a part of the Carlsbad seawater desalination project once the wetland project is completed and is fully functional. Typically it takes three to five years for a coastal wetland project to be fully functional and to begin to yield enhanced habitat and GHG sequestration benefits.

Carbon emission offsets by investing in renewable energy projects. An alternative approach to offset GHG emissions of a given desalination project is to invest in renewable energy projects located in the service area of the desalination plant. For example, the owner of the Carlsbad seawater desalination project plans to invest in a number of green power projects (rooftop photovoltaic systems, diesel bus conversion to clean-natural gas vehicles, etc.) with its public partners who will be receiving desalinated water from the Carlsbad plant. The total carbon footprint offset for the desalination plant is projected at 2,260 MWh/yr or 561 metric tons of CO_2/year (4.1 percent of net carbon footprint).

The mitigation costs of the various alternatives are summarized in Table 4-6.

Project annual net-zero carbon emission balance. Table 4-5 summarizes the total and net carbon footprint estimates of the Carlsbad seawater desalination project and quantifies GHG emission reduction and mitigation options, which are planned to be implemented in order to reduce the plant net carbon emission footprint to zero. Analysis of data presented in Table 4-5 indicates that for this example case study up to 40 percent of the GHG emissions associated with seawater desalination and drinking water delivery will be reduced by on-site reduction measures, and the remainder will be mitigated by off-site mitigation projects and purchase of renewable energy credits. It should be noted that the contribution of on-site GHG reduction activities is expected to increase over the useful life (i.e., in the next 30 years) of the project because of the following key reasons:

- In the near future, most power suppliers in the U.S. are planning to significantly increase the percentage of green power sources in their electricity supply portfolio, which in turn will reduce their Emission Factor and the net desalination plant carbon footprint.

- Advances in seawater desalination technology are expected to yield further energy savings and carbon footprint reductions. Over the last 20 years, the use of power for the production of one gallon of fresh water by seawater desalination has decreased more than two times. This trend is projected to continue in the future.

Table 4-5 Desalination project net GHG emission zero balance

Source	Total Annual Power Use (MWh/year)	Total Annual Emissions (metric tons CO_2/year)
Carbon Dioxide Emission Generation		
Seawater Desalination and Product Water Delivery—High Energy Efficiency Design	246,000	61,100
Carbon Emission Reduction Due to Reduced Water Imports	190,700	47,400
Total Net Power Use and Carbon Emissions (Item 1–Item 2)	55,300	13,700
On-site Carbon Dioxide Emission Reductions		
Energy Efficient Plant Design	Accounted for in Item 1	Accounted for in Item 1
Use of Warm Cooling Water	(12,300)	(3,100)
Green Building Design	(500)	(124)
On-site Solar Power Generation	(777)	(193)
Use of CO_2 for Water Production	NA	(2,100)
9. Reduced Energy for Water Reclamation	(1,950)	(484)
Subtotal On-site Power/GHG Emission Reduction (Sum of Items 4 through 9)	(15,527)	(6,001)
Off-site Carbon Dioxide Emission Mitigation		
CO_2 Sequestration by Re-vegetation of Wildfire Zones	(NA)	(166)
CO_2 Sequestration in Coastal Wetlands	(NA)	(304)
Investing in Renewable Energy Projects	(2,260)	(561)
Other Carbon Offset Projects and Purchase of Renewable Credits	(37,513)	(6,668)
Subtotal Off-site Power/GHG Mitigation Reduction (Sum of Items 11 through 14)	(39,773)	(7,699)
Total Net CHG Emission Balance (Item 3–Item 9–Item 14)		0

Notes: NA – not applicable. Numbers in parentheses indicate reduction.
Courtesy of Water Globe Consulting

Table 4-6 Unit costs of carbon footprint reduction alternatives

Alternative	Unit Cost (US$/metric ton CO_2 reduced)
Green Building Design	3,400
On-site Solar Power Generation	1,900
CO_2 Sequestration in Coastal Wetlands	400
CO_2 Sequestration by Revegetation of Wildfire Zones	200
Use of CO_2 for Water Production	70

Courtesy of Water Globe Consulting

The most costly carbon footprint reduction options are the green building design (US$3,400/metric ton CO_2) and the installation of rooftop solar power generation system (US$1,900/metric ton CO_2). Development of new coastal wetlands is a very promising carbon footprint reduction option (US$400/metric ton CO_2), which could be several times less costly than the construction of a solar panel generation system. Similarly, reforestation could also be a cost-competitive GHG reduction alternative (US$200/metric ton CO_2). As compared to green power generation alternatives (solar and wind power) reforestation and wetland mitigation have added environmental benefits. For example, the new coastal wetlands developed in relation to a seawater desalination project could create habitat for species that are impacted by the intake operations of the desalination plant via impingement and entrainment of these species on the intake screens.

NOISE, AIR POLLUTION, AND TRAFFIC

Noise, air pollution, and traffic associated with the construction and operation of seawater desalination plants are similar to those generated during implementation of conventional water treatment plant projects.

Of these, only noise is of specific importance because the high-pressure reverse-osmosis feed pumps operate at very high rotational speed and are usually a significant source of noise pollution. The key sources of noise at the plant are the large high-pressure pumps that feed the RO treatment trains and the interconnected energy recovery devices. These noise sources should be located in the reverse osmosis building, which would contain the generated noise.

Usually desalination plants are equipped with large intake seawater pumps, pretreatment filter transfer pumps, and product water transfer pumps, which are often located outdoors. The potential noise mitigation measures for these pumps are as follows:

1. **Use of centrifugal pumps** – Centrifugal pumps that have relatively low noise levels will be used for these applications, as an alternative to higher noise-level piston pumps.

2. **Use of water-cooled pump motors** – The main source of noise in a centrifugal pump station are the pump motors. Water-cooled motors may be used instead of standard air-cooled motors to reduce noise levels.

3. **Installation of acoustic enclosures** – Commercially available acoustic enclosures can be installed around the pump motors or the entire pump station to contain and dissipate the noise from the outdoor mechanical equipment.

4. **Installation of sound curtains** – Industrially sewn sound curtains can be installed around the pump stations using floor-mounted hardware.

5. **Installation of the pumps and motors in an enclosed building** – If required, the pumps and motors can be installed in an enclosed building designed to attenuate sound sources.

Often the noise in the main desalination building, which houses the high-pressure pumps and energy recovery devices, is attenuated by acoustic control panels installed on the walls of the building.

REFERENCES

AWWA. 1999. M46, *Reverse Osmosis and Nanofiltration*, Denver, Colo.: American Water Works Association.

AWWA Water Desalination Committee. 2004. *Water Desalination Planning Guide for Water Utilities*. Hoboken, N.J.: John Wiley & Sons, Inc.

AWWA & ASCE. 1990. *Water Treatment Plant Design, 4rd Edition*, American Waterworks Association and American Society of Civil Engineers. New York: McGraw-Hill.

Bay, S. and D. Greenstein, 1992/93. "Toxic Effects of Elevated Salinity and Desalination Waste Brine," in *Southern California Coastal Water Research Project Annual Report*. Costa Mesa, Calif. (available online at scwrp.org/pubs/annrpt/92/93).

Brevik, E.C. and J.A. Homburg. 2004. A 5000 Year Record of Carbon Sequestration from a Coastal Lagoon and Wetland Complex, Southern California, USA, *Catena*, 57:221-322.

CEC. 2005. California's Water-Energy Relationship, Final Staff Report, November 2005. (CEC-700-2005-011-SF). Sacramento, Calif.: California Energy Commission.

CCAR. 2008. Annual Emissions Report, San Diego Gas and Electric, 2006, California Climate Action Registry. (www.climateregistry.org/CARROT/Public/reports.aspx, accessed May 2008).

City of Carlsbad. 2005. Environmental Impact Report for Precise Development Plan and Desalination Plant EIR SCH# 2004041081, May.

City of Huntington Beach. 2005. Draft Recirculated Environmental Impact Report, *Seawater Desalination Project at Huntington Beach*, April.

Crisp, G. 2007. *Desalination - the focus of Australia's current membrane interest*, AWA Membranes Specialty Conference II. Melbourne, Vic.: Australian Water Association.

D.A. Lord & Associates Pty Ltd. 2005. *Ecological Assessment of the Effects of Discharge of Seawater Concentrate from the Perth Seawater Desalination Plant on Cockburn Sound*; Report No. 05/028/1. Perth, Australia.

Einav, R. and F. Lokiec. 2003. "Environmental aspects of a desalination plant in Ashkelon," *Desalination*, 156:79-85.

Fernandez-Torquemada, Y., J.L. Sanchez-Lizaso, and J.M. Gonzalez-Correa. 2005. Preliminary results of the monitoring of the brine discharge produced by the SWRO desalination plant of Alicante (SE Spain). *Desalination*, 182: 395.

Gille, D. 2003. Seawater Intakes for Desalination Plants. *Desalination*, 156: 249-256.

GWA–Government of Western Australia, Department of Environment and Conservation. 2007. *Licence for Prescribed Premises - Perth Seawater Desalination Plant. Licence Number 8108/1*; Perth, WA. 63.

Hammond, M., N. Blake, P. Hallock-Muller, M. Luther, D. Tomasko, and G. Vargo. 1998. Effects of Disposal of Seawater Desalination Discharges on Near Shore Benthic Communities, Report of Southwest Florida Water Management District and University of South Florida. Tampa, Fla.

Hunt, H.C. 1996. "*Filtered Seawater Supplies—Naturally,*" Desalination & Water Reuse, August/September, Volume 6/2.

IPCC. 2007. Climate Change 2007: Impacts, Adaptation, and Vulnerability. Contribution of Working Group II to the Third Assessment Report of the Intergovernmental Panel on Climate Change [Parry, Martin L., Canziani, Osvaldo F., Palutikof, Jean P., van der Linden, Paul J., and Hanson, Clair E. (eds.)]. Cambridge University Press, Cambridge, United Kingdom.

McPherson, G.E., K. I. Scott, J.R. Simpson, Q. Xiao, and P.J. Peper. 2000. Tree Guidelines for Coastal Southern California Communities, USDA Forest Service, Pacific Southwest Research Station (http://www.ufei.org/files/pubs/cufr_48.pdf accessed May 2008).

Mickley, M. C. 2006. "Membrane Concentrate Disposal: Practices and Regulation", Desalination and Water Purification Research and Development Program Report N. 123 (Second Edition). Washington, D.C: U.S. Department of Interior, Bureau of Reclamation.

Missimer, T.M. 1994. *Water Supply Development for Membrane Water Treatment Facilities.* Boca Raton, Fla.: Lewis Publishers/CRC Press.

Missimer, T.M. 1999. Raw Water Quality – The Critical Design Factor for Brackish Water Reverse Osmosis Treatment Facilities. *Desalination and Water Reuse*, 9/1: 41-47.

NRC. 2001. *Climate Change Science: An Analysis of Some Key Questions.* Committee of the Science of Climate Change, National Research Council (NRC). (Available online at http://books.nap.edu/html/climatechange, accessed May 2008).

Okely, P. N., J.P Antenucci, and J. Imberger. 2007. *Field Investigation of the Impact of the Perth Seawater Desalination Plant Discharge on Cockburn Sound During Summer,* Center for Water Research, University of Western Australia: Perth.

Peters, T. and D. Pinto. 2006. Sub-seabed Drains Provide Intake Plus Pretreatment. *Desalination and & Water Reuse*, 16(2): 23-27.

Purnama, A., H.H. Al-Barwani, and M. Al-Lawatia. 2003. Modeling dispersion of brine waste discharges from a coastal desalination plant. *Desalination*, 155: 41.

Rando, A.J. and C.J. Brady. 1966. *Unique Salt Water Supply System Serves Aquarium,* Public Works, August.

Rhodes, M. 2006. Marine Management is High Priority. *The International Desalination and Water Reuse Quarterly.* 16:30.

Santa Ana RWQCB. 2006. *Order No. R-8-2006-0034, NPDES No. CA 8000403, Waste Discharge Requirements for the Poseidon Resources (Surfside) LLC Seawater Desalination Facility Discharge to the Pacific Ocean*, August.

San Diego RWQCB. 2006. *Order No. R-9-2006-0065, NPDES No. CA 0109223, Waste Discharge Requirements for the Poseidon Resources Corporation Carlsbad Desalination Project, Discharge to the Pacific Ocean via the Encino Power Station Discharge Channel*, August.

Schwartz, J. 2000. *Beach Well Intakes for Small Seawater Reverse Osmosis Plants,* The Middle East Desalination Research Center, Muscat, Sultanate of Oman.

Truilio, L. 2007. Notes on Carbon Sequestration and Tidal Salt Marsh Restoration. (www.sfbayjv.org/tools/climate/CarbonWtlandsSummary_07_Trulio.pdf accessed May 2008)

Weber, C.I., W.B. Horning, D.J. Klemm, T.W. Nieheisel, P.A. Lewis, E.L. Robinson, J. Menkedick, and F. Kessler. 1998. "Short-term Methods for Estimating the Chronic Toxicity of Effluents and Receiving Waters to Marine and Estuarine Organisms". *EPA/600/4-87/028.* National Information Service, Springfield, Va.

USEPA. 2006. U.S. Climate Action Plan (available online at: www.state.gov/g/oes/rls/rpts/car/, accessed May 2008). Washington, D.C.: USEPA.

USEPA. 2008. Inventory of U.S. Greenhouse Gas Emissions and Sinks: 1990-2006. (USEPA#430-R-08-005). Washington, D.C.: USEPA.

Van Senden, D. and B.M. Miller. 2005. *Stratification and Dissolved Oxygen Issues in Cockburn Sound Pertaining to Discharge of Brine From Desalination*; Technical Report 2005/03; The University of New South Wales, Water Research Laboratory: Manly Vale, NSW.

Voutchkov, N. 2004. Thorough Study is Key to Large Beach Well Intakes, *The International Desalination and Water Reuse Quarterly*, May/June, Vol. 14/No.1.

Voutchkov, N. 2004. Seawater Desalination Costs Cut Through Power Plant Co-location, *Filtration + Separation*, September.

Voutchkov, N. 2006. Innovative Method to Evaluate Tolerance of Marine Organisms. *Desalination and Water Reuse*, 16(2): 28-34.

Watson, I.C. 2003. *Desalination Handbook for Planners*, Third Edition. Desalination and Water Purification Research Program Report No. 72. Washington, D.C.: Bureau of Reclamation.

Wetterau, G., L. Voelz, K. Klinko, R. Siminitch. 2009. Tapping the Ocean in Monterey County, California: The San City Desalination Facility, 2009 Annual Water Reuse Conference, September, Seattle, Washington.

Wilf, M. and L. Awerbuch. 2006. *Guidebook to Membrane Technology: Reverse Osmosis, Nanofiltration and Hybrid Systems Process Design, Applications and Economics*. Hopkinton, Mass.: Desalination Publications.

Wright, R.R., T.L. Missimer. 1997. *Alternative Intake Systems for Seawater Membrane Water Treatment Plants*, Proceedings of International Desalination Association, Madrid, 3: 407-422.

AWWA MANUAL M61

Chapter **5**

Cost of Treatment

Kurt Kiefer
Tom Pankratz

INTRODUCTION

During the initial stages of a desalination project, decision makers need accurate cost models that can forecast complete project costs to obtain project funding or financing. In developing these cost models, it is extremely important to consider not only the capital costs of the desalination plant itself, but all of the costs associated with financing, permitting, implementation, and operation and maintenance of the facility. It is difficult to generalize some of these costs as they are case specific; however, this chapter is intended to aid in the development of capital and operation and maintenance (O&M) cost estimates where these can be generalized. Cost parameters will be presented for key project components including source water collection, pretreatment, desalination, posttreatment, storage and distribution, and disposal of concentrate and other residuals. Generating initial capital and O&M cost estimates for these project components is an important step in the development of the total project and the final cost of the finished water reaching the customer.

SUMMARIZING PROJECT COSTS

Project costs basically fall into one of two cost categories:

- Capital costs: those that include the cost of equipment, materials, and labor to construct the desalination plant as well as costs to develop, permit, design, build, and finance the desalination project.

- Operation and maintenance (O&M) costs: those that include the cost of labor, energy, chemicals, and other consumables (RO membranes, cartridge filters, etc.) necessary for operation and maintenance of the desalination plant as well as the costs associated with residuals disposal, project administration, environmental monitoring, and permit compliance.

Typically both capital and O&M costs are evaluated to determine a total cost of water and/or to develop life cycle costs for a project. These costs are discussed in more detail later in this chapter.

Capital Costs

Capital costs include two cost components:

- Construction costs include the cost of equipment, buildings, pipelines, and other physical facilities that make up the project, including the contractor's general and administrative costs, costs of bonding and insurance, profit, and financing costs. These costs are often referred to as *hard capital costs* and may represent 60 to 80 percent of the total capital costs.

- Owner's indirect costs includes the cost of engineering, environmental, and siting issues, legal, administrative, and financial services needed to plan, design, and permit the project. These costs are often referred to as *soft costs* and may represent 20 to 40 percent of the total capital costs.

The construction cost of a desalination project, which is typically the major portion of the capital costs, depends on many factors including the following:

- Source water quality
- Product water quality goals
- Size (capacity) of the desalination project
- Intake type and distance from the plant
- Means of conveying the product water to customers
- Means of disposing of residuals (concentrate and pretreatment solids)
- Site development issues, such as,
  - Availability and cost of land
  - Geotechnical conditions
  - Site development
  - Land use
  - Architectural constraints
  - Environmental/permitting requirements
  - Other community concerns
  - Availability (and cost of) power
  - Access

A detailed analysis of key factors that have significant influence on plant costs is presented in Wilf et al. (2007). Development of a desalination project basically involves identifying a water source that can be developed to cost effectively treat and deliver water to a location where there is a demand or a need for the water. The source water quality can dictate the level of treatment and the type of desalination process required as well as the extent of pretreatment needed ahead of the desalination process. These factors can have significant impacts on the capital cost of the project. Project location and site selection can also greatly influence project costs. Site selection is often a balancing act between finding a location that results in the lowest overall cost for site development, source water intake options, pre- and posttreatment, product water storage and distribution, residuals disposal, and environmental/permitting requirements.

Operation and Maintenance Costs

A plant's operations and maintenance (O&M) cost includes the costs of actually operating and maintaining the desalination facility and producing desalinated water. O&M costs include such cost components as power, labor, chemicals, membrane replacement, concentrate disposal, repairs, and replacement parts.

O&M cost may be further broken down into two categories – fixed and variable costs. Fixed O&M costs are those costs that are independent of the amount of water treated and may range from 10 to 40 percent of the total O&M costs (excluding debt service, which is sometimes considered a fixed O&M cost).

Variable O&M costs are dependent on and vary with the amount of water that is actually treated at a given time. For example, labor costs (including wages/salaries and benefits) are usually fixed costs, whereas power and chemicals are variable costs. The energy costs for desalination plants can vary greatly depending on the salinity of the source water and type of plant, the unit power cost (electricity, oil, or other fuel), the use of energy-recovery devices, and other factors.

CONSTRUCTION COSTS

Figure 5-1 presents typical construction costs for seawater desalination plants in U.S. dollars per gallon per day installed plant capacity (2010 dollars). The seawater costs are for single-purpose plants, where desalinated water is the only product from the facility, and assume an open ocean intake with associated pretreatment facilities.

Although collocating a seawater desalination plant with a power generation or other industrial facility can result in significant savings on both the capital and operating costs for the desalination facility, environmental requirements related to industrial cooling water intakes have limited the use of collocation to date.

The costs presented in Figure 5-1 are "on-site construction" costs only. They do not include indirect (engineering, legal, financial, etc.) costs or contingencies. Neither do they include "off-site" costs for such improvements as conveying feedwater to the site, transporting product water to customers, bringing power to the site, and so on. The cost curves also assume no unusual geotechnical, architectural, environmental, etc., conditions that could significantly increase the construction cost.

ESTIMATING CAPITAL COSTS

The following sections briefly discuss the factors that affect capital costs.

Source Water Quality

Two major aspects of source water quality impact the cost of a desalting project:

- The presence of inorganic (silt, for example) and organic (algae, etc.) suspended solids

- The presence of dissolved substances, including TDS and other dissolved inorganic and organic contaminants.

The temperature of the source water is also important because as it decreases, the cost of membrane desalination tends to increase due to increased operating (feed) pressure, although this disadvantage may be partially offset by decreased salt passage and improved product quality as compared to operation at higher feedwater temperature. In some cases, where the product quality is close to the regulatory limits, operation at higher temperature may result in unacceptable product quality unless corrective action is taken in the system design, such as the use of higher rejection membranes or provision of a partial or complete second pass to meet specific quality objectives for the project. Such increased design requirements may partially offset the cost savings associated with lower pressure operation at higher temperature.

Product Water Quality Goals

In general, more stringent product water quality goals will increase the cost of desalinated water. For example, production of boiler makeup water with a membrane plant typically involves a two-pass system, which can decrease the overall system recovery, increase capital cost, and require higher operating costs due to increased power, chemical and membrane replacement costs than single-pass systems. In the case of drinking water, the minimum product water quality goals must meet the applicable drinking water regulations. However, in some cases, utilities may specify a product quality requirement for a specific parameter(s) that is more stringent than the regulatory requirement for drinking water because they may want to more closely match the quality of water that they are producing from other sources or because of other concerns such as disinfection by-product formation potential, sodium-restricted diets, or plant irrigation considerations. For these reasons, on some desalination projects, utilities have specified more stringent limits for total dissolved solids, chloride, sodium, hardness, boron, or bromide than were required by the drinking water regulations.

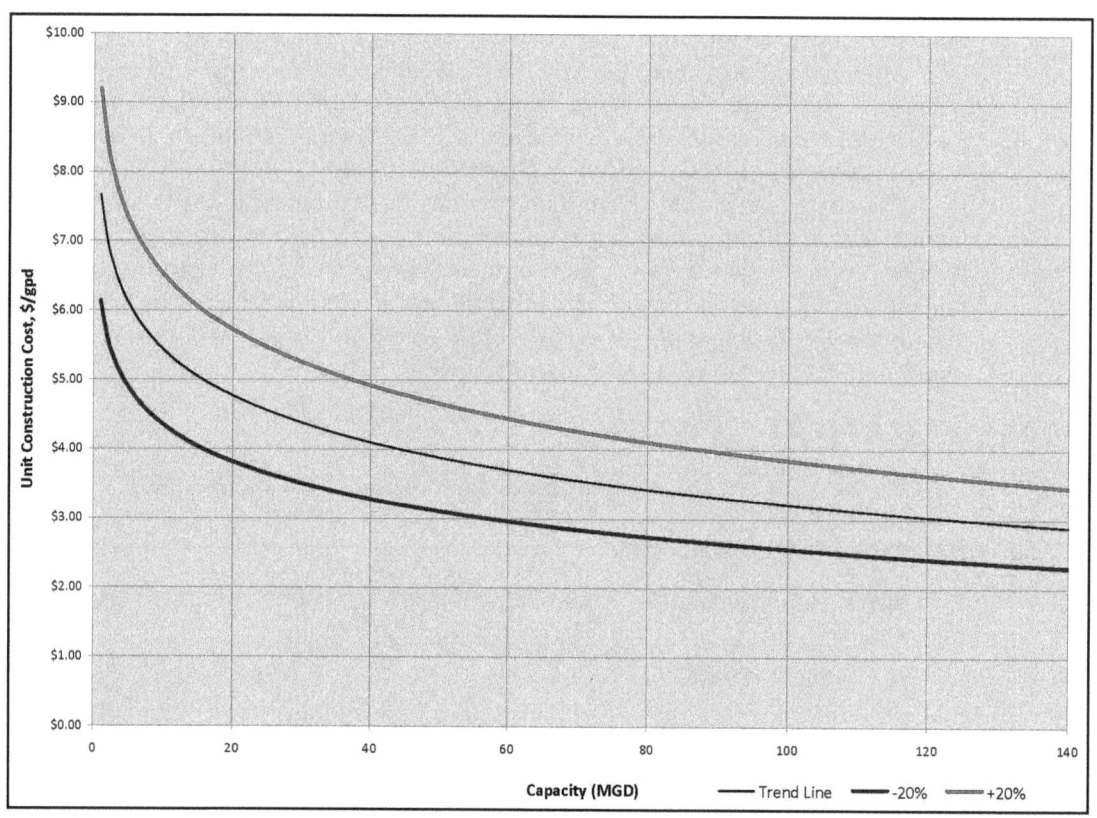

Courtesy of Tom Pankratz/Water Desalination Report

Figure 5-1 Seawater RO construction cost

Meeting these more stringent water quality goals may require the use of higher rejection membranes or the use of a partial or complete second pass, which will result in increases in capital and operating costs. For this reason, utilities developing desalination projects should closely examine the costs versus the benefits of specifying more stringent water quality objectives than required by the regulatory authorities.

Raw Water Intake

Seawater desalination facilities require a raw water intake system capable of providing a reliable quantity of seawater, preferably with consistent quality. To meet these objectives, it is essential that a thorough assessment of site conditions be conducted. Physical characteristics, meteorological, and oceanographic data, and the potential effects of pollution, navigation, and fouling by marine organisms must be evaluated.

Historically, an intake represented between 5 and 20 percent of a plant's total capital cost. However, due to increasingly stringent environmental requirements a modern intake system can represent as much as 50 percent of the capital cost of some seawater desalination facilities, as has been the case for recent facilities constructed in Australia. For the 35 mgd (132 MLD) Gold Coast facility, as an example, the reported cost of tunneling for the intake and outfall was $213 million, while the remaining facility construction costs were $532 million (Crisp 2010).

The primary environmental impact of a seawater intake relates to the impingement and entrainment of marine life. Impingement occurs when marine organisms are trapped against intake screens by the force of flowing water. Entrainment occurs when smaller organisms pass through the intake screens and into the process equipment.

A variety of intake arrangements are available to mitigate impingement and entrainment, ranging from special fine screens to submerged, offshore intakes, to infiltration systems where water is collected below the seabed. These are described in more detail in Chapter 4.

Table 5-1 shows the comparative costs for various seawater intake options for the proposed 50 mgd (139 MLD) SWRO plant in Carlsbad, California. The costs were estimated in October 2007, and based on a total feedwater of 304 mgd (1150 MLD). Although the SWRO operates at approx 45 percent recovery, the excess feedwater was used to dilute the concentrate and avoid multiport diffusers.

Product Storage and Conveyance

Along with the seawater intake, storage and conveyance of product water to customers can result in a significant project cost. The cost of these facilities depends on the onsite storage volume as well as the length, size, and materials of construction of pipelines required to deliver product water to customers. The soil conditions, land use, and population density of the terrain that the transmission main will cross also greatly influences costs.

In the South Florida Water Management District (SFWMD) study of nanofiltration and brackish water RO projects, the cost of one half-day of product storage at the rated plant production capacity and conveyance, including high service pumping to the plant site boundary, was only 5 to 10 percent of the total project cost (SFWMD 2007). In contrast, the cost of the storage and transmission system for a seawater desalination project in Southern California, in which the product water was to be pumped 20 miles (32 km) into the mountains to tie into the regional water supply was approximately equal to the cost of the desalination process including pretreatment and posttreatment (Cooley et al. 2006).

These examples illustrate that the cost of product storage and conveyance can vary considerably from project to project. For projects with storage capacity of one half-day of production capacity or less and a short transmission distance, the cost can be as low as 5 percent of total project cost. However, for large storage volumes and long conveyance distances, the cost can approach the cost of treatment.

Concentrate/Brine Disposal

A reverse osmosis plant produces two streams: a product water stream that passes through the membrane and results in a higher purity stream from which the majority of dissolved salts have been removed; and a concentrate stream that includes the remaining dissolved salts from the original feed steam (including dissolved salts removed from the product water) that remain in the water stream that does not pass through the membrane. For seawater desalination facilities, the total dissolved solids (TDS) concentration of the concentrate is usually from 1.5 to 2 times that of seawater.

Seawater desalination plants almost always return the concentrate back to the sea in such a way as to facilitate rapid dispersion and assimilation. Concentrate disposal costs are highly variable and depend on the environmental sensitivity in the area of the discharge site and the volume and water quality of the concentrate.

Simple disposal methods range from a pipe discharging into the surf zone to a complex offshore system with multiple port diffusers. The cost of a concentrate discharge system can range from 5 percent of a project's capital cost for a simple channel/pipe arrangement that discharges at the shoreline or surf zone, to more than 40 percent of the project's capital cost for a disposal system that discharges far offshore and includes a complex diffuser system.

Because of their design and construction similarities, the cost of a concentrate discharge system is often combined with the intake system cost.

Site Development

Some site development issues are straightforward. For example, it can readily be determined whether the parcel of land on which the desalination plant is to be built is of adequate size. The availability and cost of property can be somewhat easily ascertained. Whether constructing a desalination plant is compatible with surrounding land uses is also relatively easy to determine. However, it is more difficult to determine other aspects of construction costs associated with a particular property.

Consider, for instance, geotechnical suitability. One proposed seawater desalination plant was to have been sited adjacent to San Francisco Bay in an industrial area. The property already belonged to the project's owner, was undeveloped, and was large enough for the purpose. However, soil conditions indicated that a pile foundation would have been required that would have added 10 percent to the overall construction cost of the project.

Architectural constraints can also impose cost increases as compared to an industrial building design. Seawater desalination plants by their very nature are located at, or very near, the seashore. The need for desalinated seawater indicates that the surrounding area is probably developed, perhaps very densely, so locating an industrial type facility in such an area may be unacceptable. A more elaborate architectural design may be required at some additional cost to get the project permitted.

Power supply can also be a significant cost consideration. Typically, seawater desalination facilities exert large demands on the local power grid. Recently constructed seawater desalination facilities in Australia have employed renewable energy for their primary power supply. The 80 MW Emu Downs Wind Farm in Western Australia, for instance, was built in 2006 for a cost of $180 million (Australian dollars) with 66 percent of the power purchased for the 38 mgd (144 MLD) Perth desalination plant. The water treatment plant itself was built for a cost of $266 million, suggesting that the power supply for the treatment plant was 45 percent of the cost of the plant. Such power supply costs are generally not included in the reported capital costs of new desalination facilities but are rather incorporated into the O&M costs for purchased power.

Indirect Costs

Indirect costs typically include engineering, environmental permitting, legal, financial, and administrative services. For design-bid-build projects, these costs typically fall to the owner, and bid prices for these projects do not typically include indirect costs; however, utilities should carefully consider these costs in developing a budget for a desalination project. For design-build-operate and finance projects, the majority of these costs will be the responsibility of the project developer and would be included in the cost of water for the project. However, the water purchaser may also incur costs for engineering, legal, environmental permitting, and administrative costs that are not reflected in the cost of water. Each project is unique and, therefore, the magnitude of the indirect costs as a percentage of the construction cost can vary over a very wide range, depending on the size and complexity of the project.

Some percentage of the estimated construction cost is usually assumed for indirect costs when initially planning a desalination project. Typically, indirect costs used in the planning stages of a project are on the order of 30 to 40 percent of the total project construction cost.

Contingency Allowance

Project contingency is generally a function of the level of engineering design that has been completed and the level of detail of the cost estimate. In the initial planning stages of a desalination project, there will be a number of unknowns – things that have not yet been identified. Even for knowns, such as the desalination plant itself, unexpected costs will most likely arise as the project becomes more defined. Therefore, it is necessary to include a contingency allowance in the capital cost budget for the project. At the beginning of the project, the contingency allowance is relatively large. As planning and design proceed, the contingency allowance may be reduced. However, it may be that some of the contingency allowance originally included in the preliminary project budget has been reallocated to construction costs or indirect expenses as these costs were further defined.

The contingency allowance is typically expressed as a percentage of the estimated construction cost, and in the early stages of planning, the contingency allowance may be 25 percent or more of the estimated construction cost. After the final plans and specifications have been prepared, the contingency allowance may be on the order of 10 percent. Even after construction bids are received for the project, a contingency allowance is necessary to cover the unexpected costs that arise on essentially all construction projects.

ESTIMATING OPERATION AND MAINTENANCE COSTS

With respect to O&M costs, labor is typically considered a fixed O&M cost. Wage levels may change with time, but labor costs are not directly impacted by the amount of water that is desalinated. Another fixed O&M cost is RO membrane replacement.

Generally, once water has passed through a membrane, it is considered to have begun its useful life. In general a membrane is not considered to age more rapidly when it is in operation than when it is in a standby mode after it has been placed into service. For example if a membrane plant is operated only 50 percent of the time, it is not generally assumed that the membrane's life would be twice as long as it would be if the plant was operated 100 percent of the time.

In addition, some allowance should be included for repairs and replacement parts as a fixed O&M cost for planning purposes; an allowance of 2 to 3 percent of the plant construction cost is typically included to cover this. There may be other fixed O&M costs specific to a given project. Fixed O&M costs are usually evaluated on an annual cost basis.

Variable O&M costs are those costs that directly depend on the amount of water that is desalinated. Variable costs may include power, chemicals, concentrate disposal, and other miscellaneous costs. The total annual O&M cost is the sum of the fixed and variable O&M costs in dollars per year.

Fixed O&M Costs

Labor and labor overhead. With regard to labor, the cost is dependent on a number of factors, including the following:

- *The size (capacity) of the plant* – A small plant may operate with only one or two permanent staff with unattended operation during some periods of the day. In contrast, larger plants typically require operators in attendance at all times to keep up with routine maintenance, record keeping, etc.

- *The complexity of the plant* – If the treatment process is complicated, it may be necessary to have a larger operating staff to cover all the skills needed for plant operation.

- *Regulatory agency requirements* – Local or state regulatory agencies may have minimum staffing requirements for a desalination plant.

- *Owner's policy* – Some desalination plant owners may desire to have at least one operator at the plant around the clock, while another owner may wish to have as small an operating staff as possible with the plant operating unattended at times.

- *Local wages / benefits* – Some areas, particularly in or near densely populated metropolitan areas, may have higher wage/benefit costs than more rural areas.

The cost of membrane replacement is usually regarded as a fixed cost because once a membrane is placed in operation, it is generally assumed that membrane degradation begins. A membrane replacement fund may be established to accumulate money to replace membranes as needed. The two primary factors considered in establishing a membrane replacement fund are:

- The total cost of the membranes
- The expected average life of the membranes

Determining the initial cost of the membranes depends on several factors, including:

- The capacity of the desalination process
- The number of treatment passes required
- The recovery expected
- The membrane design flux

These factors can be ascertained with reasonable certainty and are usually part of a construction cost estimate.

Estimating the annual membrane replacement cost is basically a function of the number of membranes installed, the projected membrane replacement cost, and the anticipated useful life of the membranes. The useful life of the membranes primarily depends on the feedwater quality and the effectiveness of pretreatment. For this reason, the life expectancy of second pass membranes is typically longer than first pass membranes. With an increasing number of seawater reverse osmosis plants installed, a larger database of typical membrane replacement rates for various pretreatment scenarios is being created.

Membrane life expectancy has been increasing from a traditional life expectancy of three years to the point that life expectancies of five to seven years are not unusual, especially with MF/UF as pretreatment.

An allowance for repairs and replacement parts beyond membrane replacement should also be included when planning a desalination project. The cost for repairs and replacement parts is typically taken as 2 to 3 percent of the plant construction cost.

Variable O&M Costs

Power. All of the power required by the desalination project should be included in estimating the cost. For instance, the power required to pump the feedwater to the plant from the source(s) of supply and the product water to customers should be accounted for as well as the power required for the desalination process itself. The power required (kW of demand and kWh consumed) can usually be relatively accurately defined in the preliminary phases of project planning.

It is important to obtain as accurate an estimate of power costs ($/kWh) as possible because power can be a substantial portion of the O&M cost. This is especially true for seawater desalination plants because they consume more power than brackish water treatment facilities. Most of the power cost will be a variable cost. However, some small percentage will be a fixed cost, because even when the desalination system is not operating or is operating at reduced capacity, there will be some power demand for building loads and site lighting.

Chemicals. The types and amounts of chemicals that might be required in a desalination plant can vary significantly. Chemicals are required for the following purposes:

- *Filtration* – If particle removal is needed ahead of the desalination process, chemicals may be necessary to increase filter performance. Such chemicals may include coagulants, filter aids such as polymers, acid for pH adjustment, and disinfection chemicals to control biogrowth.

- *Desalination process feedwater* – Historically it has been common practice to add a scale inhibitor and/or acid ahead of the SWRO membranes to inhibit calcium carbonate scale formation and reduce the risk of scaling/fouling. The Stiff-Davis Index is typically used as an indicator of the calcium carbonate scaling potential. More recently, however, for SWRO systems with TDS in the range of 35,000 mg/l and pH in the range of 8.0 or less have been successfully operated without acid and antiscalant addition in the recovery range of 45 to 50 percent. With higher efficiency energy recovery systems, designers sometimes opt for slightly lower recoveries in the 45 to 47 percent range to allow some additional safety margin with respect to scaling potential. Operation at higher recoveries will still likely require the addition of acid and/or scale inhibitor to avoid scale formation.

- *Posttreatment* – Seawater RO plants typically reduce hardness and alkalinity to very low levels, which results in a relatively high corrosion potential of RO permeate. As a result, it is common practice to add hardness and alkalinity to stabilize the RO permeate. Lime (calcium hydroxide) or calcite (calcium carbonate) is typically added to increase the hardness of the finished water. Acid or carbon dioxide is also added to react with the lime or calcite and produce alkalinity. A disinfection residual in the form of chlorine or chloramines is also added after desalination. The selection of the disinfectant type is often based on the disinfection practices used for water from other sources that may be blended with finished water from the desalination facility.

- *Membrane cleaning* – RO membranes require periodic cleaning where the frequency of cleaning and the type of chemicals required vary depending on the particular case. Generally, a low pH (acid) solution is used to clean mineral scales and high pH (caustic soda) solution is used to clean biological fouling. In addition, a detergent may be used from time to time. If membrane filtration is used to filter the desalination process feedwater, the same chemicals may be used to clean the pretreatment membranes. There are also a variety of proprietary cleaning products available for more challenging cleaning applications.

The cost for chemicals depends on the quality and quantity of the water being treated and the product water quality goals. Chemical consumption data from other SWRO plants treating seawater of similar quality can be used to estimate chemical treatment needs and costs for a new desalination facility. However, if such data are not available, pilot testing is typically performed to better quantify chemical requirements and subsequent costs, particularly for very large SWRO systems.

Residuals disposal. Desalination plant residuals include the concentrate stream and residual solids that were removed from the raw water prior to membrane treatment. Residual solids include suspended solids along with coagulation chemicals that were used in the pretreatment process. For any desalination process, it is important to develop a strategy for concentrate volume reduction and disposal and to determine the associated costs and permitting requirements in the planning stages of the project.

For seawater desalination projects, concentrate is usually returned to the ocean. If sources for dilution of the concentrate are available, such as cooling water from a power plant or ocean discharges from a wastewater treatment plant, dilution of the concentrate flow by mixing with these streams can facilitate the permitting process and reduce the cost of permitting. In addition, use of existing outfall structures can significantly reduce the cost of concentrate disposal. Where dilution alternatives are not readily available, installation of diffusers can be used to promote rapid mixing and dispersion of the concentrate. Diffusers can reduce discharge concentrations close to background ocean salt concentrations within relatively short distances from the discharge points. The costs associated with additional pressure drops across the diffusers to promote mixing, and the energy cost associated with discharging concentrate farther away from the plant site and/or further off shore to take advantage of prevailing ocean currents to further promote mixing, should be considered in the O&M costs. Costs for monitoring water quality in the discharge area to assure compliance with permitting requirements is also an important factor to account for in the O&M costs.

The cost of residuals disposal from the pretreatment system can vary widely depending on raw water quality and the options for disposal of residual solids. For installations where suspended solids levels are relatively low and where membrane filtration can be used without coagulants or low dosing of coagulants, it may be possible to return residuals to the ocean with minimal treatment costs. On the other hand, if the desalination plant is close to an estuary where high runoff flows and high turbidity can be experienced, the amount of solids may be too great for permitable discharge. If beneficial uses for these residuals cannot be found, disposal of these solids in a landfill may be required. Disposal costs for landfills can vary significantly from location to location, and in some areas disposal cost may be high and can become a significant O&M cost. There are also treatment costs associated with dewatering (filter press or centrifuge) that should be considered in the development of O&M cost estimates. For these reasons residuals management should be carefully evaluated in the development of a desalination project.

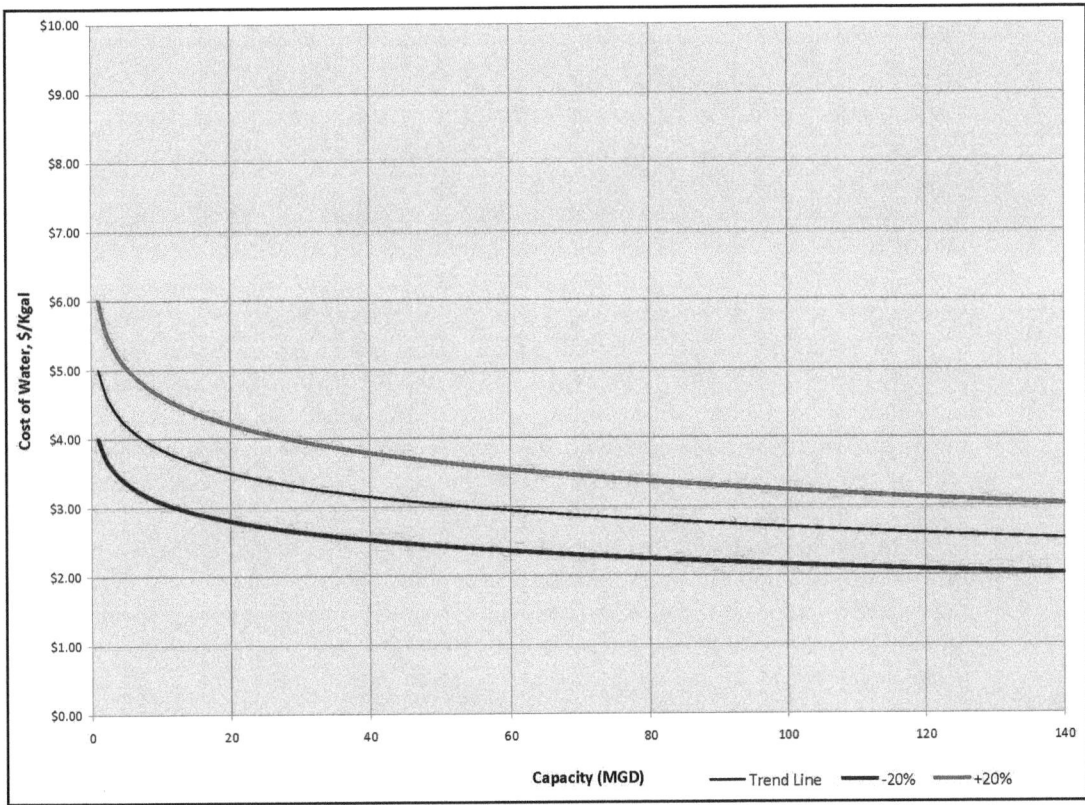

Courtesy of Tom Pankratz/Water Desalination Report

Figure 5-2   Seawater RO cost of water

FINANCING COST

In most cases, desalination projects are financed using borrowed money, and interest payments, which can be significant, will increase the total cost of the project. The following example illustrates the potential significance of the cost of borrowing on the total cost of a desalination project. Begin by assuming the following:

- A 20 mgd (76 MLD) SWRO project is being constructed
- Capital cost = $120,000,000
- The desalination plant produces 95 percent of its design annual yield (20 mgd × 365 days/year × 95% = 6,935,000 kgal/yr at rated capacity)
- Financing is obtained for the project with repayment terms of

    Term of the loan = 25 years   Interest rate = 8%

Assuming compounding annually, the annual payment would be $11,241,500, and the impact of the financing cost on the cost of water would be $1.62/kgal ($0.43/m^3). The cost of financing would be approximately $161,000,000 over the life of the project.

COST OF WATER

The sum of the annual capital repayment and annual O&M cost divided by the volume of water produced during the year is termed *the cost of water* and is usually expressed as dollars per unit volume, for example, in dollars per thousand gallons ($/kgal), dollars per cubic meter ($/m^3), or dollars per acre-foot ($/AF).

Total production costs are dependent on the specific site and application, desalination technology used, plant facilities constructed, power, labor, and chemical costs, annual water production, and many other factors. Figure 5-2 presents typical treated water costs for SWRO desalination plants (2010 dollars). Typical treated water costs for a 5-mgd (19 MLD) seawater desalination plant range from approximately $3.10/kgal to $4.75/kgal ($0.82/m^3 to $1.25/m^3); recent cost estimates for very large plants (50 mgd or larger) indicate that costs may range from $3.00/kgal to $3.60/kgal ($0.79/m^3 to $0.95/m^3). Depending on the treatment plant capacity, the cost of water for SWRO desalination facilities typically ranges from approximately $2 to $5 per thousand gallons or $0.5 to $1.3 per cubic meter (WDR 2008).

As stated previously, caution should be used when extrapolating the cost data presented here to a specific current or future project. These costs are presented only to give the reader an approximate idea of costs and detailed estimates should be made as part of any desalination plant feasibility analysis. Special consideration should be given to the anticipated cost of power, which may account for 50 to 70 percent of the annual O&M costs for a seawater desalination facility.

SUMMARY

As discussed in the capital cost section, the installation of a new desalination facility most likely includes the development of other related infrastructure including the seawater intake, raw water transmission main, pretreatment and posttreatment systems, desalination, product storage and distribution, site improvements, residuals management, and connection of the facility to the power grid. Careful planning is required to ensure that all the costs associated with the development of a new desalination plant are included during planning stages and project budget development.

Tables 5-2 (estimated capital cost), 5-3 (estimated O&M costs), and 5-4 (estimated water costs) present an example of how a preliminary cost estimate for a desalination project might be presented. The example is based on:

- Desalination of seawater from an open ocean intake
- Product water design capacity = 50 mgd (189 MLD)
- Annual product water production rate based on:
  - 100% of rated plant capacity = 18,250 million gallons per year (mgy) or 69,000 m^3/yr
  - Average day demand at 80% of rated capacity = 14,600 mgy (55,000 m^3/yr)
- Desalination process recovery = 50%
- Single pass design

Table 5-3 illustrates how O&M costs for this desalination plant might be presented. Two annual production rates are provided to show the impact on the cost of water of spreading the fixed costs of O&M over a greater volume of water production. As shown in the table, the variable O&M costs were adjusted to reflect the fact that if more product water is produced, the annual costs for power and chemicals will increase correspondingly.

Annual fixed costs do not change. The result is that the unit O&M cost per thousand gallons (kgal) of water produced at the annual production rate of 18,250 mgy is lower than the unit O&M cost for producing only 14,600 mgy from this facility.

Table 5-4 compares the cost of water for the same annual production rates. The annual capital repayment cost is constant as are the fixed O&M costs. Only the variable O&M costs change with the change in annual production rate. As the total cost of water figures in the table show, producing more water from the desalination plant reduces the cost of water.

The annual capital cost was calculated using an interest rate of 8 percent and a 25-year amortization period. (Annual debt payment = $0.0937 \times \$198,200,000$).

Table 5-1 Seawater intake alternatives cost example

Alternative	Cost ($Million)
Use Existing Power Plant Intake	$0
New Open Seawater Intake	$150
New Slant Well Sub-Sea Intake	$415
New Horizontal 'Raney Well' Sub-Sea Intake	$438
New Sub-Seabed Infiltration Gallery	$638
New Vertical Beachwell Intake	$650

Courtesy of Tom Pankratz/Water Desalination Report

Estimated cost of alternative seawater intakes for Poseidon's proposed 50 mgd (189 MLD) SWRO plant in Carlsbad, California, October 2007.

Table 5-2 Seawater RO plant capital cost example

Plant Capacity: 50 mgd (189 MLD)	2010 Dollars
Cost item	
Direct capital (construction) costs	
1. Site preparation, roads and parking	1,982,000
2. Intake	9,910,000
3. Pretreatment	15,856,000
4. RO system equipment	71,352,000
5. Post-treatment	3,964,000
6. Concentrate disposal	5,946,000
7. Residuals Management	2,973,000
8. Electrical System	4,955,000
9. Product Storage & Distribution	3,964,000
10. Buildings	9,910,000
11. Startup Costs	3,964,000
Subtotal direct (construction) costs	$134,776,000
Project engineering services	$21,802,000
Project Development Costs	
1. Administration, contracting and management	4,955,000
2. Environmental permitting	9,910,000
3. Legal services	2,973,000
Subtotal project development	$17,838,000
Project financing costs	$23,784,000
Contingency	12,883,000
Subtotal indirect capital costs	$63,424,000
Total capital costs	$198,200,000
Unit Cost, $/gpd	$3.964
Unit Cost, $/m^3/d	$1,047

Courtesy of Tom Pankratz/Water Desalination Report

COST OF TREATMENT 97

Table 5-3   Annual Operation and Maintenance Cost Example Treatment Technology: SWRO

Assumptions Unit power cost = $0.10 per kW/hr		2010 Dollars
Item No.	Description	Plant Capacity (mgd) 50
Fixed		
1.	Operation and maintenance labor	$761,000
2.	Replacement parts and materials	$5,945,000
3.	Replacement membranes	$2,536,000
Variable Costs		
4.	Power	$22,638,000
5.	Chemicals	$6,042,000
Annual Production at Rated Capacity, (mgy)		18,250
Annual O&M Cost at Rated Capacity		**$37,922,000**
Unit Cost at Rated Capacity, $/kgal		$2.08
Unit Cost at Rated Capacity, $/m^3		$0.55
Average Day Capacity/Rated Capacity		80%
Annual Production at Avg Day Demand, (mgy)		14,600
Annual O&M Cost at ADD Capacity		**$32,186,000**
Unit Cost at Rated Capacity, $/kgal		$2.20
Unit Cost at Rated Capacity, $/m^3		$0.58

Courtesy of Kurt Kiefer

Table 5-4   Annual Cost of Water Example Treatment Technology: SWRO (with a power plant)

Assumptions Unit power cost = $0.10 per kW/hr		2010 Dollars
Item No.	Description	Plant Capacity (mgd) 50
Production Costs at Rated Capacity		
1.	Equivalent annual capital cost	$18,564,000
2.	Annual O&M Cost - Fixed	$9,242,000
3.	Annual O&M Cost - Variable	$28,680,000
Total Annual Cost:		$56,486,000
	Annual Production at Rated Capacity, (mgy)	18,250
	Annual Production Cost at Rated Cap. ($/kgal):	**$3.09**
	Annual Production Cost at Rated Cap. ($/m^3)	**$0.82**
	Average Day Capacity/Rated Capacity	80%
1.	Equivalent annual capital cost	$18,564,000
2.	Annual O&M Cost – Fixed	$9,242,000
3.	Annual O&M Cost – Variable	$22,944,000
Total Annual Cost:		$50,750,000
Annual finished water production rate (mgy)(3):		14,600
Annual Production Cost at ADD ($/kgal):		**$3.48**
Annual Production Cost at ADD ($/m^3)		**$0.92**

Courtesy of Kurt Kiefer

REFERENCES

AWWA. 2004. *Water Desalting Planning Guide for Water Utilities.* Hoboken, N.J.: John Wiley & Sons, Inc.

AWWA. 2007. M46 – *Reverse Osmosis and Nanofiltration.* Denver, Colo.: American Water Works Association.

Cooley, H., P. Gleick and G. Wolff. 2006. *Desalination, With a Grain of Salt: A California Perspective.* Oakland, Calif.: Pacific Institute.

Crisp, G. 2010. *Reducing Carbon Footprint of a Seawater Desalination Plant – An Australian Case Study,* Multi-State Salinity Coalition Annual Salinity Summit.

Pankratz, T. 2008. "Water Desalination Report." 44(33).

South Florida Water Management District. 2007. *Water Supply Cost Estimation Study.* Tampa, Fla.

Watson, I.C., O.J. Morin, Jr., and L. Henthorne. 2003. *Desalting Handbook for Planners 3rd edition.* Washington, D.C.: U.S. Bureau of Reclamation.

Wilf, M., L. Awerbouch, C. Bartels, M. Mickley, G. Pearce and N. Voutchkov. 2007. *The Guidebook to Membrane Desalination Technology.* L'Aqula, Italy: Balaban Desalination Publications.

AWWA MANUAL M61

Chapter 6

Safety and Security

Gary Silverman
Irving Moch

SAFETY

Introduction

Access to sufficient quantities of safe water for drinking and domestic uses and also for commercial and industrial applications is critical to health and well-being, and the opportunity to achieve human and economic development. Desalination plants treat water infused with salts (seawater and brackish water) or other contaminants as their sources. It is imperative that these plants operate in a safe manner with respect to the people who ingest its water, the plant operators, the environment, and the processing equipment.

The following sections discuss the safety of the various items, components, materials, and features encountered in a desalination plant. Membrane and thermal facilities have somewhat similar listings. The main differences are the replacement of the membranes, pressure vessels, and pretreatment sections of membrane plants with that of the vaporization chamber, recirculation loops, and heat exchangers for thermal desalination.

System Component Design

System equipment components should be designed to have a relatively useful life (15 to 30 years), to the extent possible and practical. All plant components (equipment and structures) should be designed such that failures are repairable. Suppliers should provide documentation on individual equipment components verifying that the components are suitable for the water quality stated in the specifications.

Expected useful life should be detailed in the system specifications for the various mechanical and electrical equipment and devices. The system supplier, fabricator, and individual component manufacturer should provide documentation that these components meet the requirements. All system design and verification documentation should be summarized by the system supplier in the Operation and Maintenance manual.

Products and Components

Materials of construction. The structures and equipment components of a desalination system should be constructed of materials that are corrosion resistant to the environment in which they are placed. High quality stainless steel and plastics may be acceptable materials of construction provided they can withstand the rigors of their internal and external environments.

Pressure ratings. Many products and components may be used in systems where pressures are other than ambient. In such instances, the materials of construction must meet the codes established by federal, state, and local regulatory agencies and professional associations such as ASME International.

Temperature variations. Thermal systems and sometimes membranes operate under temperature conditions other than ambient. All materials used in such conditions must be structurally unaffected by these temperature variations at maximum operating pressures.

Toxicity levels. All products and components must not be toxic to the public or the environment in their intended use. Where the water is for potable use, systems should be certified under NSF International Standards 60 and 61 and other standards promulgated by NSF International, AWWA, and ASTM International. The products and components must be inert to the environment when discarded, meeting regulatory standards.

Other considerations. All products and components must be safe to use. Protection of personnel, equipment and the environment must be assured at all times through the proper monitoring and use with installation of relief valves and indicating, recording and de/activation instrumentation.

General Plant Safety Components

- Lifting lugs
- Anchor bolts
- Pressure vessel straps for restricting movement
- Identification labels and numbering on equipment
- Color identification of electrical and flow piping with directional markings if needed
- Instrumentation test equipment
- Pressure vessel probing equipment.

Readily Accessible Plant Procedural Manuals

- Operating instructions
- Maintenance instructions
- Equipment specifications
- Raw material specifications
- Cost records for control
- Appropriate shift log books
- Safety manual
- Emergency evacuation procedures
- Up-to-date equipment and P&I drawings

Safety Stations
- Showers
- Eye wash stands
- Containment dunes for flooding potential
- Stocked first aid stations

Safety Equipment
- Safety shoes and eyeglasses
- Hard hats
- Masks and respirators
- Clothing for handling hazardous chemicals and thermal conditions

Interface Points
- Interface points during construction coordinated with the contractor and subcontractors as applicable
- Piping: interfaces for all connecting piping shown on drawings
- Support structure, anchors, and mounting and leveling devices available
- All electrical conduits and wiring terminated in the panel board
- An input/output (I/O) control panel provided
- All conduits, cables, and wiring from instruments, valves, and other control devices terminated in the panel board
- All pneumatic piping to valves and other devices provided and terminate to a single point for each type of air supply

Execution
Installation. The various components should be installed per the specification and contract. The system supplier should have a competent field representative on site at least at the following milestones:

- Initial delivery of major equipment
- Beginning of the installation of the equipment, particularly membrane elements
- Final connections and flushing of all equipment
- Other times as specified
- Membranes should only be installed after a thorough flushing of all piping to remove construction debris
- All flushing water should be clean, potable water and not allowed to sit in piping for extended periods

Start-up and commissioning. Start-up and commissioning should be the responsibility of the party responsible for overall project implementation. The utility or owner, and the equipment suppliers should determine, in advance, which party will be responsible for operation of the system.

Start-up should include electrical and mechanical checks of all equipment, leak checking, flushing, placing units in service, reconfirming the function aspects of the system, flow and performance verification, etc. For a membrane plant, it should only occur after the quality of the feed water to the SWRO system has met specifications i.e., pretreatment step has been activated and is functioning properly.

Training. The system supplier should provide training to the owner's maintenance and operating staff. This should consist of both classroom and hands-on activities. Training is a continuing activity. Existing staff should have scheduled periodic reviews and updates of all operating and maintenance procedures. These procedures should be in written and electronic form, and readily available to personnel at all times. It is important that new operational and maintenance employees have training before they are certified to work on process equipment, instrumentation, and/or electrical supply.

Safety field testing. Field testing should include mechanical and electrical testing, which, at a minimum, should encompass equipment operation (pumps, valves, etc.), instrument verification, leak testing, power and signal connection verification, and PLC/PC program verification. A functional test of all operating equipment should be performed to verify that the equipment is performing safely.

SECURITY

A seawater desalination plant, like any other water treatment facility, is a critical asset of a water utility. Its function is to provide a reliable source of safe drinking water to the public. As such, steps should be taken to prevent any breach in security that may serve to compromise the facility's infrastructure or contaminate its product water. This can be done by anticipating undesired acts and implementing a plan to prevent them from occurring.

In May 2009, AWWA Standard G430-09, entitled *Security Practices for Operation and Maintenance*, became effective. This standard provides important guidance to utilities in implementing security measures. As stated, its purpose is to "define the minimum requirements for a protective security program for a water or wastewater utility that will promote the protection of employee safety, public health, public safety, and public confidence."

The following is a brief summary of AWWA G430-09. Readers of this manual are encouraged to consult the standard directly for more detailed security guidance.

Commitment to Security

An effective security program begins with a visible commitment to security by senior leadership of the utility. Leaders should establish a security culture by promoting security awareness throughout the organization. This may be accomplished by providing opportunities for employee suggestions; implementing training programs; issuing ID badges; incorporating security into job descriptions; measuring progress; and rewarding employees for positive behavior in enhancing security awareness. Specific roles and responsibilities should be assigned for creating, maintaining and implementing the security plan. Security should have a sustained focus within the utility by devoting appropriate resources to the topic in terms of budget and staff time.

Risk Assessment

A risk or vulnerability assessment should be conducted for the seawater desalination plant. In the case of a proposed facility, the assessment should be done proactively so that the findings of the assessment can be incorporated into the design and construction. The assessment should include, at a minimum, a characterization of the facility; an identification and prioritization of the adverse consequences to avoid; a determination of the critical assets that may be targeted; an assessment of the likelihood of a malevolent act occurring;

and an evaluation of appropriate countermeasures. The resultant security plan should be reviewed and updated periodically as well as following a significant event, such as after new construction, new information about a potential threat, or after having sustained an attack or breach of security. The security plan should be integrated with other operational plans, including emergency response and business continuity plans, so that its priorities are considered and addressed in a broad context. These plans should be tested regularly through training, drills, or simulations.

Access Control and Intrusion Detection

The security plan will most certainly require implementing various means of access control and intrusion detection to protect critical assets, including physical structures and information technology (IT) and SCADA systems. These will be accomplished through installation of physical components and systems and through implementing various policies and procedures.

Physical Measures

Physical measures may include hardening walls and incorporating intrusion prevention devices on windows and doors; erecting fences and other barriers; providing locking devices for doors, gates, and hatches; and installing monitoring systems with intrusion alarms. The latter category may utilize motion detectors, enhanced lighting or closed-circuit TV surveillance, among others. Surveillance methods should be implemented to detect chemical, biological, or radiological contamination through on-line monitoring systems and laboratory testing. This can be done directly or through surrogate monitoring. In the case of IT and SCADA systems, physical protections may include installing and maintaining firewalls; separating business and operational systems; installing a system for virus protection; securing SCADA equipment locations; and incorporating encryption technologies.

Policies and Procedures

Policies and procedures should be adopted to control access to critical assets by personnel through a hierarchical access card system or other means to maintain security. Staff should be required to display identification at all times, and visitor access should be controlled through a sign-in and escort system. A procedure for deliveries should include chain-of-custody control or tamper-evident packaging requirements. When legally permitted, background checks should be made on employees, and a protocol should be instituted to immediately rescind access privileges for employees who have been terminated or resigned.

Communications

Strong internal and external communication strategies should be developed and implemented relative to security issues. Regular and ongoing communication with employees is critical to ensure that security matters are taken seriously, to promote employee safety during an event, and to enable effective employee participation during an event. Effective communications with response organizations, regulatory agencies, and customers is also vital. Collaborative partnerships should be established with key agencies to ensure cooperation and effective coordination during emergency response and recovery. Mutual aid and assistance agreements should be reached with neighboring utilities and other relevant agencies.

Index

NOTE: *f.* indicates a figure; *t.* indicates a table.

Affordable Desalination Collaboration (ADC), 42
Aggressiveness index (AI), 21
Algal toxins, 18
Ashkelon (Israel) desalination plant, 59, 60*f.*,
 and open discharge of concentrate, 61
AWWA Standard G430-09, *Security Practices for Operation and Maintenance*, 102

Boron, 18, 21–22
 concentration of, and feedwater temperature, 34, 35*f.*
 and irrigation and agricultural water use, 21–22, 23
 toxicity, 20*f.*, 22
Bromide, 20
 brominated DBPs, 20–21

Calcium carbonate, 37
Calcium carbonate precipitation potential (CCPP), 21
Calcium hydroxide (hydrated lime), 37
Calcium oxide (quicklime), 37
Capacitive deionization (CDI), 12
Capital costs, 83
 and concentrate disposal, 88
 construction (hard capital) costs, 84–85, 86*f.*
 contingency allowance, 89
 defined, 83
 estimate (example), 94, 96*t.*
 indirect (soft) costs, 84, 89
 and product storage and conveyance, 87–88
 and product water quality goals, 86–87
 for site development, 88
 and source water intake, 87, 95*t.*
 and source water quality, 85–86
Carbon dioxide, 37
Carboneras (Spain) Seawater Desalination Plant, 67
 and colocation, 67, 67*f.*
Carlsbad (California) Seawater Desalination Plant, 65, 73*f.*
 Climate Action Plan, 73–79, 78*t.*
 costs for intake alternatives, 87, 95*f.*
Cartridge filtration (in pretreatment), 29, 30*t.*
CCPP. *See* Calcium carbonate precipitation potential
CDI. *See* Capacitive deionization
Chemical costs, 91–92
Chloride, 18
 and industrial water use, 22–23
 and irrigation and agricultural water use, 21–22, 23
 toxicity, 20*f.*, 22
Chlorination (in pretreatment), 29, 30*t.*
Coagulation, flocculation, and clarification (in pretreatment), 29, 30*t.*

Concentrate discharge
 capital costs, 88
 disposal costs, 92
Concentrate discharge (environmental impacts and mitigation), 57–58
 beneficial use of concentrate, 69
 coastal well disposal, 68–69
 concentrate disposal methods, 59–69, 60*t.*
 deep well injection, 68
 discharge through existing power plant outfall (colocation), 65–67, 66*f.*, 67*f.*
 discharge through existing wastewater treatment plant outfall, 63–65, 64*f.*
 discharge through new ocean outfall, 59–63, 60*f.*
 mechanisms of concentrate impact, 58
 and osmotic conformers, 58–59
 and osmotic regulators, 59
 Perth (Australia) ocean outfall discharge, 57, 61–63, 61*f.*, 62*f.*
 quantity as function of plant recovery, 57
 and salinity tolerance threshold of organisms, 58–59
Corrosion, 19, 21, 43
 crevice corrosion, 43, 44*f.*
 galvanic corrosion, 43–45, 44*t.*
 general mitigation practices, 45–47
 microbial corrosion, 45
 pitting corrosion, 45, 46*t.*
 pitting resistance equivalent number (PREN), 45, 46*t.*
 stress corrosion, 45
 types and mitigation, 43–45
 under-deposit corrosion, 45
Cost of water, 93*f.*, 94
 estimate (example), 95, 97*t.*
Costs. *See* Capital costs; Cost of water; Financing costs; Operation and maintenance costs
Cryptosporidium, 19, 19*t.*
 and log reduction credits for various treatment processes, 36, 37*t.*

Dechlorination (in pretreatment), 29, 30*t.*
Desalination of seawater, 1–2
 developing technologies, 9–13
 large-scale facilities, 1–2
 membrane-based technologies, 3–6
 operational facilities (US), 1–2, 2*t.*
 thermal technologies, 6–9
 worldwide growth of capacity, 1, 2*f.*
Developing technologies
 capacitive deionization, 12
 forward osmosis, 9–11, 10*f.*

freeze/thaw, 12
 membrane distillation, 11
 supercritical desalination, 12–13
Disinfection, 36
 log reduction credits for various treatment processes, 36, 37t.
Disinfection by-products (DBPs), 19–21
Dissolved air flotation (DAF), in pretreatment, 29, 31t.
Domoic acid, 18
Dual work exchangers (DWEER), 40
 flow diagram, 40, 41f.
 multi-train approach, 40, 42f.
 typical installation, 40, 41f.

E. coli, 19
Electrodialysis (ED), 3, 6
 as electric potential driven process, 3
Electrodialysis reversal (EDR), 3, 6
 as electric potential driven process, 3
Energy recovery, 38, 42
 and Affordable Desalination Collaboration (organization), 42
 centrifugal devices, 38, 39f.
 devices compared, 42, 43t.
 dual work exchangers (DWEER), 40, 41f., 42f.
 Francis Turbines, 38
 hydraulic turbochargers, 38, 39f.
 Pelton impulse turbines (PIT), 38, 39f.
 positive displacement devices, 38–40, 40f., 41f., 42f.
 pressure exchangers, 38, 40f.
Environmental impacts, 49–50
 assessing in comparison with supply alternatives, 50
 of concentrate discharge, 57–69
 of greenhouse gas emissions, 72–79
 of noise, 79
 pretreatment residuals management, 69–72
 of source water intakes, 50–57
Environmental Protection Agency (EPA). *See* U.S. Environmental Protection Agency

Filtration (in pretreatment), 29, 30t.
Financing costs, 93
Forward osmosis (FO), 9–11, 10f.
 as concentration-driven process, 3
Francis Turbines, 38
Freeze/thaw, 12

Giardia, 19, 19t.
 and log reduction credits for various treatment processes, 36, 37t.
Greenhouse gas emissions (environmental impacts and mitigation), 72
 assessing gross carbon footprint, 73–74
 carbon dioxide, 72
 carbon dioxide sequestration in coastal wetlands, 77
 carbon emission offsets by investing in renewable energy projects, 77
 Climate Action Plans, 73–79, 78t.
 fluorinated gases, 72
 green building design, 75–76
 management approaches, 73–79
 methane, 72
 nitrous oxide, 72
 offsetting carbon footprint by reduced water imports, 74–75
 principal greenhouse gases, 72
 project annual net-zero carbon emission balance, 77–79, 78t.
 unit costs of carbon footprint reduction alternatives, 77, 78t.
 use of carbon dioxide for water production, 76–77

Hydraulic turbochargers, 38, 39f.

Langelier Saturation Index (LSI), 21

MED. *See* Multiple effect distillation
Membrane distillation (MD), 11
Membrane-based technologies, 3
 and concentration gradient, 3
 driving forces in, 3
 and electric potential, 3
 electrodialysis, 6
 electrodialysis reversal, 6
 nanofiltration, 5–6
 and pressure, 3
 reverse osmosis, 3–5, 4f.
Membranes
 aging of, 89
 replacement costs, 90–91
Micro-sand enhanced clarification (MES), in pretreatment, 29, 31t.
Microfiltration (MF), in pretreatment, 29, 31t.
Multiple effect distillation (MED), 7–9, 8f.
Multistage flash distillation (MSF), 7, 8f.

Nanofiltration (NF), 3, 5–6
 as pressure driven process, 3
 two-pass systems, 35–36
Nephelometric turbidity units (ntu), 17
Noise (environmental impacts and mitigation), 79

Operation and maintenance costs, 83, 85
 for chemicals, 91–92
 defined, 83–84
 estimate (example), 94–95, 97t.
 for filtration chemicals, 91
 fixed, 85, 89, 90–91
 for labor, 89, 90
 and membrane age, 89
 for membrane cleaning chemicals, 92
 for membrane replacement, 90–91
 for posttreatment chemicals, 91
 for power, 91
 for process feedwater chemicals, 91
 for repairs and parts, 89, 91
 for residuals disposal, 92
 variable, 85, 90, 91–92

Pelton impulse turbines (PIT), 38, 39f.
Perth (Australia) Seawater Desalination Plant, 57, 62–63
    discharge configuration (with diffuser), 61, 61f.
    mixing zone, 61, 62f.
Posttreatment (remineralization), 37
Power
    consumption as function of recovery, 34, 34f.
    costs, 91
Practical Salinity Units (psu), 15
Pressure exchangers, 38, 40f.
Pretreatment residuals management, 69
    disposal costs, 92
    residuals from granular media filters and membrane pretreatment systems compared, 70–72, 71t.
    spent backwash water, 69–70
    types of residuals, 69, 70t.
Pretreatment, 27
    cartridge filtration, 29, 30t.
    chemical dosage for scale inhibition, 29, 30t.
    chlorination, 29, 30t.
    coagulation, flocculation, clarification, 29, 30t.
    dechlorination, 29, 30t.
    dissolved air flotation (DAF), 29, 31t.
    filtration, 29, 30t.
    microfiltration (MF), 29, 31t.
    micro-sand enhanced clarification (MES), 29, 31t.
    at recently installed SWRO plants, 29–31, 32t.
    and Silt Density Index, 28
    for subterranean (beach well) intake facilities, 28–29
    in surface supply intake facilities, 29, 30t., 31t.
    technologies, 28–32
    ultrafiltration (UF), 29, 31t.
    upflow solids contact clarification, 29, 31t.
    and water quality, 27–28
Product water
    capital costs for storage and conveyance, 87–88
    stability, 21
Product water quality, 17
    aesthetic concerns, 23–24
    and algal toxins, 18
    and boron, 18, 21–22
    and chloride, 18, 21–23
    and corrosion, 19, 21
    and disinfection by-products (DBPs), 19–21
    in distribution systems, 19–21
    goals for, and capital costs, 86–87
    health concerns, 17–21
    in industrial use, 22–23
    in irrigation and agricultural use, 21–22, 23
    mineral content, 17–18
    nonregulated parameters, 18
    pathogen removal, 19, 19t.
    and sodium, 17, 21–22, 23
    stability, 21
    taste and odor concerns, 24
    temperature, 23

Residuals management. *See* Concentrate discharge; Pretreatment residuals management
Reverse osmosis (RO), 3
    basic concept, 3–5, 4f.
    cellulose acetate (CA) membranes, 4
    as pressure driven process, 3
    thin-film composite (TFC) membranes, 4
    See also Seawater reverse osmosis

Safety, 99
    and construction materials, 100
    equipment, 101
    field testing for, 102
    general plant components for, 100
    and installation of plant equipment, 101
    and interface points, 101
    and O&M training, 102
    and pressure ratings, 100
    and procedural manuals, 100
    in start-up and commissioning, 101–102
    stations, 101
    and system component design, 99–100
    and temperature variations, 100
    and toxicity levels, 100
Salinity tolerance threshold, 58–59
Santa Barbara (California) Seawater Desalination Plant, 63, 64f.
    concentrate discharge through existing wastewater treatment plant outfall, 63–64
Saxitoxin, 18
Scale inhibition by chemical addition (in pretreatment), 29, 30t.
SCD. *See* Supercritical desalination
SDI. *See* Silt Density Index
Seawater
    mineral concentrations, 15, 16t.
    salinities (worldwide), 15, 16f.
    See also Source water quality
Seawater reverse osmosis (SWRO)
    basic concept of RO, 3–5, 4f.
    cellulose acetate (CA) membranes, 4
    design parameters, 33–36
    feedwater temperature, 34, 35f.
    membrane flux, 33
    power consumption as function of recovery, 34, 34f.
    pretreatment, 27–32
    recovery, 33
    thin-film composite (TFC) membranes, 4
    two-pass nanofiltration, 35–36
    two-pass systems in removal of boron, sodium, and chloride, 23
Security, 102
    access control, 103
    commitment to, 102
    communications, 103
    intrusion detection, 103
    physical measures, 103
    policies and procedures, 103
    risk (vulnerability) assessment, 102–103

Security Practices for Operation and Maintenance, AWWA Standard G430-09, 102
Silt Density Index (SDI), 28
Sodium, 17, 21–22, 23
Sodium hydroxide, 37
Source water intakes
    capital costs, 87, 95*t*.
    pretreatment for subterranean (beach well) facilities, 28–29
    pretreatment in surface supply facilities, 29, 30*t*., 31*t*.
Source water intakes (environmental impacts and mitigation), 50
    alternative open intake technologies, 55, 56*t*.
    beach well intakes, 51–54, 52*f*., 53*f*.
    coastal wetland habitat and groundwater), 54
    impingement and entrainment of organisms, 50–51, 54–56, 56*t*.
    loss of coastal habitat during construction, 54
    open intake construction, 57
    open ocean intakes, 54–56, 56*t*.
    subsurface intake construction, 54
    subsurface intakes, 51
Source water quality, 15
    capital costs, 85–86
    mineral concentrations (seawater/freshwater comparison), 15, 16*t*.
    salinity, 15, 16*f*.
    total organic carbon, 17
    turbidity, 17
Supercritical desalination (SCD), 12–13

Tampa Bay (Florida) Seawater Desalination Plant, 66
    and colocation, 66, 66*f*.
Thermal evaporation technologies, 6–7
    multiple effect distillation, 7–9, 8*f*.
    multistage flash distillation, 7, 8*f*.
    vapor compression, 9, 9*f*.
Trihalomethanes (THMs), 19
Two-pass systems
    nanofiltration, 35–36
    in removal of boron, sodium, and chloride, 23

U.S. Environmental Protection Agency (EPA), 17
Ultrafiltration (UF), in pretreatment, 29, 31*t*.
Upflow solids contact clarification (in pretreatment), 29, 31*t*.

Vapor compression (VC), 9, 9*f*.
Viruses, and log reduction credits for various treatment processes, 36, 37*t*.

Water quality. *See* Product water quality; Source water quality

AWWA Manuals

M1, *Principles of Water Rates, Fees, and Charges*, Fifth Edition, 2000, #30001PA

M2, *Instrumentation and Control*, Third Edition, 2001, #30002PA

M3, *Safety Practices for Water Utilities*, Sixth Edition, 2002, #30003PA

M4, *Water Fluoridation Principles and Practices*, Fifth Edition, 2004, #30004PA

M5, *Water Utility Management*, Second Edition, 2004, #30005PA

M6, *Water Meters—Selection, Installation, Testing, and Maintenance*, Second Edition, 1999, #30006PA

M7, *Problem Organisms in Water: Identification and Treatment*, Third Edition, 2004, #30007PA

M9, *Concrete Pressure Pipe*, Third Edition, 2008, #30009PA

M11, *Steel Pipe—A Guide for Design and Installation*, Fifth Edition, 2004, #30011PA

M12, *Simplified Procedures for Water Examination*, Fifth Edition, 2002, #30012PA

M14, *Recommended Practice for Backflow Prevention and Cross-Connection Control*, Third Edition, 2003, #30014PA

M17, *Installation, Field Testing, and Maintenance of Fire Hydrants*, Fourth Edition, 2006, #30017PA

M19, *Emergency Planning for Water Utility Management*, Fourth Edition, 2001, #30019PA 254 ductile -iron pipe and fitings

M20, *Water Chlorination/Chloramination Practices and Principles*, Second Edition, 2006, #30020PA

M21, *Groundwater*, Third Edition, 2003, #30021PA

M22, *Sizing Water Service Lines and Meters*, Second Edition, 2004, #30022PA

M23, *PVC Pipe—Design and Installation*, Second Edition, 2003, #30023PA

M24, *Dual Water Systems*, Third Edition, 2009, #30024PA

M25, *Flexible-Membrane Covers and Linings for Potable-Water Reservoirs*, Third Edition, 2000, #30025PA

M27, *External Corrosion—Introduction to Chemistry and Control*, Second Edition, 2004, #30027PA

M28, *Rehabilitation of Water Mains*, Second Edition, 2001, #30028PA

M29, *Fundamentals of Water Utility Capital Financing*, Third Edition, 2008, #30029PA

M30, *Precoat Filtration*, Second Edition, 1995, #30030PA

M31, *Distribution System Requirements for Fire Protection*, Fourth Edition, 2008, #30031PA

M32, *Distribution Network Analysis for Water Utilities*, Second Edition, 2005, #30032PA

M33, *Flowmeters in Water Supply*, Second Edition, 2006, #30033PA

M36, *Water Audits and Loss Control Programs*, Third Edition, 2009, #30036PA

M37, *Operational Control of Coagulation and Filtration Processes*, Third Edition, 2011, #30037PA

M38, *Electrodialysis and Electrodialysis Reversal*, First Edition, 1995, #30038PA

M41, *Ductile-Iron Pipe and Fittings*, Third Edition, 2009, #30041PA

M42, *Steel Water-Storage Tanks*, First Edition, 1998, #30042PA

M44, *Distribution Valves: Selection, Installation, Field Testing, and Maintenance*, Second Edition, 2006, #30044PA

M45, *Fiberglass Pipe Design*, Second Edition, 2005, #30045PA

M46, *Reverse Osmosis and Nanofiltration*, Second Edition, 2007, #30046PA

M47, *Capital Project Delivery*, Second Edition, 2010, #30047PA

M48, *Waterborne Pathogens*, Second Edition, 2006, #30048PA

M49, *Butterfly Valves: Torque, Head Loss, and Cavitation Analysis*, First Edition, 2001, #30049PA

M50, *Water Resources Planning*, Second Edition, 2007, #30050PA

M51, *Air-Release, Air/Vacuum, and Combination Air Valves*, First Edition, 2001, #30051PA

M52, *Water Conservation Programs—A Planning Manual*, First Edition, 2006, #30052PA

M53, *Microfiltration and Ultrafiltration Membranes for Drinking Water*, First Edition, 2005, #30053PA

M54, *Developing Rates for Small Systems*, First Edition, 2004, #30054PA

M55, *PE Pipe—Design and Installation*, First Edition, 2006, #30055PA

M56, *Fundamentals and Control of Nitrification in Chloraminated Drinking Water Distribution Systems*, First Edition, 2006, #30056PA

M57, *Algae: Source to Treatment*, First Edition, 2010, #30057PA

M58, *Internal Corrosion Control in Water Distribution Systems*, First Edition, 2011, #30058PA